# Proceedings

Ein stetig steigender Fundus an Informationen ist heute notwendig, um die immer komplexer werdende Technik heutiger Kraftfahrzeuge zu verstehen. Funktionen, Arbeitsweise, Komponenten und Systeme entwickeln sich rasant. In immer schnelleren Zyklen verbreitet sich aktuelles Wissen gerade aus Konferenzen, Tagungen und Symposien in die Fachwelt. Den raschen Zugriff auf diese Informationen bietet diese Reihe Proceedings, die sich zur Aufgabe gestellt hat, das zum Verständnis topaktueller Technik rund um das Automobil erforderliche spezielle Wissen in der Systematik aus Konferenzen und Tagungen zusammen zu stellen und als Buch in Springer.com wie auch elektronisch in Springer Link und Springer Professional bereit zu stellen. Die Reihe wendet sich an Fahrzeug- und Motoreningenieure sowie Studierende, die aktuelles Fachwissen im Zusammenhang mit Fragestellungen ihres Arbeitsfeldes suchen. Professoren und Dozenten an Universitäten und Hochschulen mit Schwerpunkt Kraftfahrzeug- und Motorentechnik finden hier die Zusammenstellung von Veranstaltungen, die sie selber nicht besuchen konnten. Gutachtern, Forschern und Entwicklungsingenieuren in der Automobil- und Zulieferindustrie sowie Dienstleistern können die Proceedings wertvolle Antworten auf topaktuelle Fragen geben.

Today, a steadily growing store of information is called for in order to understand the increasingly complex technologies used in modern automobiles. Functions, modes of operation, components and systems are rapidly evolving, while at the same time the latest expertise is disseminated directly from conferences, congresses and symposia to the professional world in ever-faster cycles. This series of proceedings offers rapid access to this information, gathering the specific knowledge needed to keep up with cutting-edge advances in automotive technologies, employing the same systematic approach used at conferences and congresses and presenting it in print (available at Springer.com) and electronic (at Springer Link and Springer Professional) formats. The series addresses the needs of automotive engineers, motor design engineers and students looking for the latest expertise in connection with key questions in their field, while professors and instructors working in the areas of automotive and motor design engineering will also find summaries of industry events they weren't able to attend. The proceedings also offer valuable answers to the topical questions that concern assessors, researchers and developmental engineers in the automotive and supplier industry, as well as service providers.

Alexander Heintzel
(Hrsg.)

# Antriebe und Energiesysteme von morgen 2022

Band 1: Antriebe

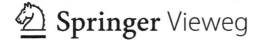

*Hrsg.*
Alexander Heintzel
Springer Fachmedien Wiesbaden
Wiesbaden, Deutschland

ISSN 2198-7432           ISSN 2198-7440    (electronic)
Proceedings
ISBN 978-3-658-41434-4       ISBN 978-3-658-41435-1    (eBook)
https://doi.org/10.1007/978-3-658-41435-1

Die Deutsche Nationalbibliothek verzeichnet diese Publikation in der Deutschen Nationalbibliografie; detaillierte bibliografische Daten sind im Internet über http://dnb.d-nb.de abrufbar.

© Der/die Herausgeber bzw. der/die Autor(en), exklusiv lizenziert an Springer Fachmedien Wiesbaden GmbH, ein Teil von Springer Nature 2023
Das Werk einschließlich aller seiner Teile ist urheberrechtlich geschützt. Jede Verwertung, die nicht ausdrücklich vom Urheberrechtsgesetz zugelassen ist, bedarf der vorherigen Zustimmung des Verlags. Das gilt insbesondere für Vervielfältigungen, Bearbeitungen, Übersetzungen, Mikroverfilmungen und die Einspeicherung und Verarbeitung in elektronischen Systemen.
Die Wiedergabe von allgemein beschreibenden Bezeichnungen, Marken, Unternehmensnamen etc. in diesem Werk bedeutet nicht, dass diese frei durch jedermann benutzt werden dürfen. Die Berechtigung zur Benutzung unterliegt, auch ohne gesonderten Hinweis hierzu, den Regeln des Markenrechts. Die Rechte des jeweiligen Zeicheninhabers sind zu beachten.
Der Verlag, die Autoren und die Herausgeber gehen davon aus, dass die Angaben und Informationen in diesem Werk zum Zeitpunkt der Veröffentlichung vollständig und korrekt sind. Weder der Verlag noch die Autoren oder die Herausgeber übernehmen, ausdrücklich oder implizit, Gewähr für den Inhalt des Werkes, etwaige Fehler oder Äußerungen. Der Verlag bleibt im Hinblick auf geografische Zuordnungen und Gebietsbezeichnungen in veröffentlichten Karten und Institutionsadressen neutral.

Verantwortlich im Verlag: Markus Braun

Springer Vieweg ist ein Imprint der eingetragenen Gesellschaft Springer Fachmedien Wiesbaden GmbH und ist ein Teil von Springer Nature.
Die Anschrift der Gesellschaft ist: Abraham-Lincoln-Str. 46, 65189 Wiesbaden, Germany

# Foreword

## Welcome

The defossilization of energy systems and transport toward renewable electricity-based solutions is the core challenge in climate protection. Fully electric and electrified drives are at the heart of future mobility. This can only be successfully implemented in close cooperation between the automotive and energy industries and policymakers.

A future global powertrain mix must systemically address the challenge of a sustainable and secure supply and distribution of energy. What is needed is an approach that is open to all technologies and considers energy resources and converters throughout the entire process. Only by shaping the technology discussion as a "not only, but also" approach towards next-generation engines in a sustainable energy system will it be possible to leverage the potentials for a targeted reduction in global $CO_2$. The perfect interaction between traffic concepts, vehicles with sustainable powertrains, and infrastructure is the prerequisite for market success.

We look forward to your participation in the hybrid event 16th International MTZ Congress on Future Powertrains.

On behalf of the Scientific Advisory Board
Prof. Dr. Peter Gutzmer
Editor-in-Charge ATZ|MTZ Group
Springer Nature

# Geleitwort

## Herzlich willkommen

Die Defossilisierung der Energiesysteme und des Verkehrs hin zu regenerativ strombasierten Lösungen ist die Kernherausforderung beim Klimaschutz. Vollelektrische und elektrifizierte Antriebe sind Kern der Zukunftsmobilität. Erfolgreich umgesetzt werden kann das nur in enger Zusammenarbeit von Automobil- und Energieindustrie sowie der Politik.

Ein zukünftiger globaler Antriebsmix muss eine nachhaltige und sichere Versorgung und Distribution von Energie systemisch berücksichtigen. Nötig ist eine technologieoffene Betrachtung von Energieträgern und -wandlern über die gesamte Wirkungskette. Nur indem die Technologiediskussion als ein „Sowohl-als-auch" in Richtung Next-Generation Engines in einem nachhaltigen Energiesystem gestaltet wird, lassen sich die Potenziale für eine nachhaltige globale $CO_2$-Senkung gezielt heben. Ein perfektes Zusammenspiel von Verkehrskonzepten, Fahrzeugen mit zukunftsfähigen Antrieben und Infrastruktur ist dabei Voraussetzung für den Markterfolg.

Wir freuen uns auf Ihre Teilnahme am Hybrid-Event 16. Internationaler MTZ-Kongress Zukunftsantriebe.

Für den Wissenschaftlichen Beirat
Prof. Dr. Peter Gutzmer
Herausgeber ATZ|MTZ-Gruppe
Springer Nature

# Inhaltsverzeichnis

Direct Drive System to Make In-Wheel Electric Vehicles
Closer to a Production Reality .................................... 1
   *Akeshi Takahashi, Makoto Ito, Tetsuya Suto, Ryuichiro Iwano,*
   *and Takafumi Hara*

$CO_2$- Life Cycle Assessment for the Porsche Taycan .................... 12
   *Otmar Bitsche and Benjamin Passenberg*

ZF eConnect: Efficient Solutions for AWD BEV ..................... 21
   *Alessio Paone, Stephan Demmerer, Matthias Winkel, Martin Ruider,*
   *Philip Endres, and Uwe Großgebauer*

Balancing of Efficiency, Costs and $CO_2$-Footprint for Future Mobility ........ 33
   *Christoph Danzer, Alexander Poppitz, Tobias Voigt, Manfred Prüger,*
   *and Marc Sens*

Requirement and Potential Analysis of Load Profile Prediction Algorithms .... 46
   *Lukas Schäfers, Pascal Knappe, Rene Savelsberg, Matthias Thewes,*
   *Simon Gottorf, and Stefan Pischinger*

Depending on Lithium and Cobalt – The Impact of Current Battery
Technology and Future Alternatives ................................ 62
   *Mareike Schmalz, Christian Lensch-Franzen, Jürgen Geisler,*
   *Amalia Wagner, Thomas Rempel, and Johannes Hüther*

System Optimization for 800 V e-drive Systems in Automotive
Applications ..................................................... 73
   *Joao Bonifacio, Felix Prauße, Michael Sperber, Thomas Schupp,*
   *Wolfgang Häge, and Viktor Warth*

FCTRAC and $BioH_2$Modul – A Way to Zero Emission Mobility
in Agriculture ................................................... 86
   *Veronica Gubin, Christian Varlese, Florian Benedikt, Johannes Konrad,*
   *Stefan Müller, Daniel Cenk Rosenfeld, and Peter Hofmann*

Systems Engineering for Fuel Cell Vehicles: From Simulation
to Prototype ..................................................... 105
   *Daniel Ritzberger, Alexander Schenk, and Falko Berg*

x    Inhaltsverzeichnis

Sustainability Assessment of an Integrated Value Chain for the Production
of eFuels . . . . . . . . . . . . . . . . . . . . . . . . . . . . . . . . . . . . . . . . . . . . . . . . . .    119
   *Jana Späthe, Manuel Andresh, and Andreas Patyk*

Creating and Sustaining User Engagement in Bidirectional Charging . . . . . . . .    130
   *Franziska Kellerer, Johanna Zimmermann, Sebastian Hirsch,*
   *and Stefan Mang*

The Energy Transition in Germany: Carbon Neutrality in the Balancing
Act between Energy Demand and Energy Supply . . . . . . . . . . . . . . . . . . . . . . .    140
   *Matthias Böger and Klaus Fuoss*

**Autorenverzeichnis** . . . . . . . . . . . . . . . . . . . . . . . . . . . . . . . . . . . . . . . . . . .    159

# Direct Drive System to Make In-Wheel Electric Vehicles Closer to a Production Reality

Akeshi Takahashi[✉], Makoto Ito, Tetsuya Suto, Ryuichiro Iwano, and Takafumi Hara

Research & Development Group, Hitachi, Ltd, Hitachi,Japan
{akeshi.takahashi.hc,makoto.ito.wg,
tetsuya.suto.rp}@hitachi.com

**Abstract.** This paper presents a compact, lightweight direct-drive system for the electric vehicle (EV) segment, which combines the motor, inverter, and brake into a single unit. This enables the entire system to be installed into the wheel and more expansive interiors and battery installation spaces, thus moving the world one step closer to a zero-emissions society. The new motor directly transmits the high driving force necessary to run an EV to the wheels, and its lightweight design and 2.5 kW/kg power density minimize the significant weight increase typically associated with in-wheel units. Moreover, implementing an in-wheel unit does not require a substantial change to the existing configuration of the suspension and other components. Driveshafts and other indirect mechanisms have been eliminated, enabling motor power to be applied directly to EV operation. This reduces energy loss by 30% and increases the range on a single charge compared with existing EVs.

**Keywords:** In-wheel motor · Direct-drive · Power density

## 1 Introduction

There is growing investment and technological development towards realizing a decarbonized society. In the motor vehicle sector, there is an especially strong legislative push for transitioning from gasoline-driven vehicles to electric vehicles (EVs).

Figure 1 shows projected global EV sales by scenario in 2020–2030 [1]. EVs include battery electric vehicles (BEVs), plug-in hybrid electric vehicles (PHEVs), and fuel cell electric vehicle (FCEVs). There are two scenarios forecasted: a) stated policies scenario and b) sustainable development scenario. The annual sales are projected to be 30–40 million units in 2030. Moreover, level 4 or 5 autonomous driving is expected to become more widespread after 2030s.

It is important to forecast the value that vehicles will provide when both electrification and autonomous driving become widespread in the future. Vehicles

---

© Der/die Autor(en), exklusiv lizenziert an Springer Fachmedien Wiesbaden GmbH, ein Teil von Springer Nature 2023
A. Heintzel (Hrsg.): ATZLive 2022, Proceedings, S. 1–11, 2023.
https://doi.org/10.1007/978-3-658-41435-1_1

should have a spacious interior that can produce a moving experience and not be restricted by cruising range. Therefore, we aim to develop an in-wheel system to expand the space inside the vehicle and improve the cruising range.

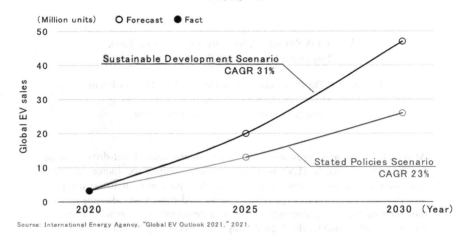

**Fig. 1.** Global EV sales by scenario, 2020–2030

In conventional EVs, the drive system is placed on the chassis, which limits the interior and/or battery space as shown in Fig. 2(a) [2, 3]. One possible solution is fitting the motor inside a wheel, but this increases the weight of the wheel and requires sweeping changes to the existing brake and suspension components as shown in Fig. 2(b) [4].

Our developed direct-drive system – drawing on Hitachi Group's broad technology and product development in the mobility sector including railways and elevators – combines the motor, inverter, and brake into a single in-wheel unit for EVs [5]. Our proposed in-wheel system is based on three main concepts as shown in Fig. 3: 1) minimal size and weight, 2) compact integration, and 3) energy conservation.

Section 2 describes in detail our motivation for developing the direct-drive system. Sections 3 and 4 introduce the key technologies developed on the basis of the three concepts. Section 5 presents the demonstration tests of the prototype and results.

Direct Drive System to Make In-Wheel ... 3

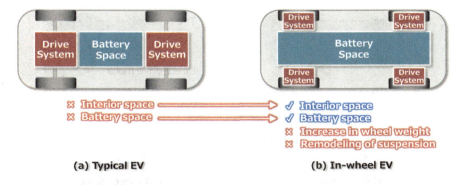

Fig. 2. Differences between typical EV and in-wheel EV

Fig. 3. Key concepts of developed in-wheel system

## 2 Configuration of In-Wheel System

### 2.1 Mechanical Structure

The mechanical structure of the in-wheel system is classified into two categories, gear drive and direct drive, as shown in Fig. 4. The gear drive shown in Fig. 4(a) is primarily made up of a high-speed motor and reduction gears. The advantage of the motor is that it can be miniaturized. However, high-speed motors tend to have a small diameter and long shaft length to attain high power because it needs to keep the centrifugal force applied to its rotor small. As a result, the space inside the wheel cannot be effectively used. Furthermore, a low-power motor must be implemented to suppress the shaft length because the reduction gears must be arranged in the axial direction of the motor. Therefore, it has been difficult to improve the power density of gear-drive structures for in-wheel systems.

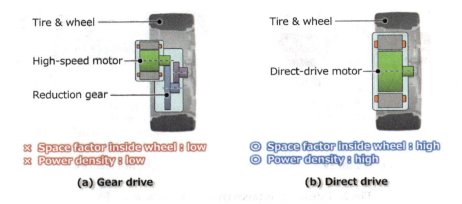

**Fig. 4.** Mechanical configurations of in-wheel systems

The direct-drive motor shown in Fig. 4(b) can have a large diameter because it rotates at a low speed, and the shaft length can be easily reduced. In addition, reduction gear is not required, and high power can be attained by increasing the space factor of the motor in the wheel. As a result, the power density can also be improved.

## 2.2 Magnetic Structure

Figure 5 shows the magnetic structures of the gear drive and the direct drive. The outer diameter of the high-speed motor in the gear drive is generally set at a small value of about $\phi 200$, as shown in Fig. 5(a), because the motor needs to keep the centrifugal force small. There are typically about eight poles, the rotation is about 15,000 min$^{-1}$, and the corresponding electric frequency is 1000 Hz. Generally, the fewer the poles in a motor, the greater the magnetic flux generated per pole pair. Therefore, the radial width of the stator and rotor needs to be increased to pass a large amount of magnetic flux, which makes it difficult to enhance the power density of the in-wheel motor.

The direct drive doubles the outer diameter of the motor to about $\phi 400$ as shown in Fig. 5(b). The number of poles is about ten times that of the gear drive, but the rotation is 1/10, so the corresponding electrical frequency is the same as that of the gear drive (1000 Hz). The amount of magnetic flux generated per pole pair is reduced to 1/10 because the number of poles has increased ten times. Therefore, the radial width of the stator and rotor can be reduced, and the power density can be improved.

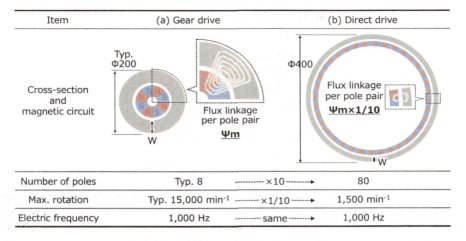

Fig. 5. Magnetic configurations of in-wheel motor

## 3 Minimal Size and Weight

Increasing the number of magnetic poles can improve a motor's driving force; however, it reduces the proportion of magnetic flux that can be effectively used and necessitates more coil weld points and welding space.

We increased the effective magnetic flux of each magnetic pole and improved driving force by placing the magnets in a Halbach array as shown in Fig. 6. The weight of the motor was reduced by using innovative flat coils to create a high-density array of coils. These key technologies have made it possible to attain 2.5 kW/kg motor power density, which is the ratio of motor output to weight including the motor housing and the driveshafts. This minimizes the in-wheel weight increase of the EV and avoids the increased energy consumption characteristic of conventional, heavier in-wheel systems.

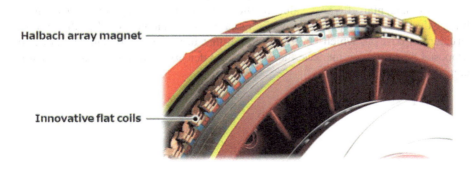

Fig. 6. Small and lightweight technology to suppress weight increase of wheel portion

**Table 1.** Specifications of in-wheel system

| Item | Value |
|---|---|
| Wheel size | 19 inch |
| Max. output | 60 kW |
| Max. torque | 960 Nm |
| Max. rotation | 1200 min$^{-1}$ |
| Supply voltage | 420 Vdc |
| Max. current | 280 Arms |
| Coolant | Oil |

Table 1 summarizes the specifications of the proposed in-wheel system. The target wheel size is 19 inches, and total output power of 4-wheel drive is 240 kW, i.e., 60 kW per wheel, which is expected to cover typical sport utility vehicles (SUVs). The maximum torque for short-term rating is 960 Nm, and the maximum rotation of the motor is 1200 min$^{-1}$, which is equivalent to a vehicle speed of about 150 km/h. The inverter for the in-wheel system is supplied with DC 420 V and outputs 280 A at maximum. The inverter and the motor are both cooled by oil, which is described in Sect. 4.

The following section describes the features of the two key technologies, the Halbach array magnet and the innovative flat coils.

### 3.1 Halbach Array Magnet

Figure 7(a) shows a comparison between the conventional and proposed Halbach array magnets. In general, Halbach array magnets have a magnetic array in which the orientation of the N pole of each magnet is rotated 90 deg. to create high-density magnetic flux at each of the motor's magnetic poles as shown in Fig. 7(a). However, the magnetic flux density of the gap cannot be sufficiently increased in the conventional structure because the magnet itself has large magnetic resistance. In our proposed Halbach array, we arranged the combination of flat-plate Nd-Fe-B magnets and allocated a core piece at the center of the pole as shown in Fig. 7(b).

The finite element analysis (FEA) results of the flux line chart indicate that the spatial-fundamental effective value of the conventional air-gap flux density is 6% higher than that of the proposed, while the amount of magnets used in the proposed structure is 0.61 pu, which is 39% less than that in the conventional. Therefore, the proposed structure can increase the air-gap flux density with fewer magnets than the conventional structure.

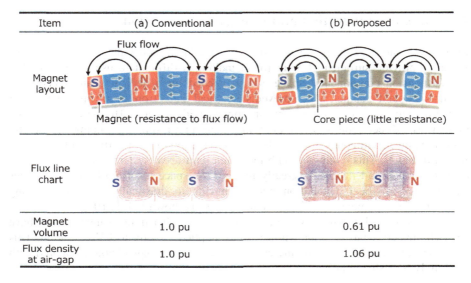

Fig. 7. Comparison of conventional and proposed Halbach array magnets

## 3.2 Innovative Flat Coils

Figure 8 shows a comparison of conventional and proposed coil configurations. As shown in Fig. 8(a), high-speed motors often implement the conventional coil configuration, or hairpin coils, with distributed winding, which enables large current density in the cross section of the coil while increasing the axial length of the coil end.

We have developed flat coils for concentrated winding to resolve the issue of the coil end. As shown in Fig. 8(b), the high-density array of the flat coils can decrease to 1/3 the axial length of the coil ends.

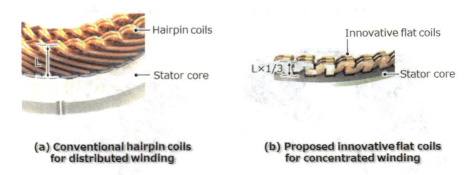

Fig. 8. Comparison of conventional and proposed coil configurations

## 4 Compact Integration and Energy Conservation

Prior EV motors have low power density and monopolize the space inside the wheel to provide sufficient driving force, making it difficult to use existing brakes and suspension components. Space is also needed for a dedicated and electrically insulated coolant channel, which prevents electrical faults from occurring when power semiconductors in the inverter come in contact with coolant. The newly developed system uses direct-cooling technology, in which high-insulating cooling oil directly cools the power semiconductors and is cycled to the motor to directly cool the coils as shown in Fig. 9. This combined with the single-unit drive system – which integrates a motor, brake, and inverter – greatly reduces the space taken up by cooling pipes and enables in-wheel installation without having to drastically alter the existing configuration of suspension and other components as shown in Fig. 10.

In the proposed in-wheel system, driveshafts and other indirect mechanisms have been eliminated, enabling motor power to be applied directly to EV operation. This reduces energy loss by 30% and increases the range on a single charge compared with existing EVs as shown in Fig. 11.

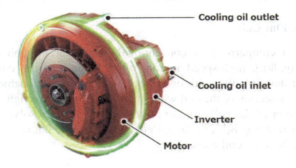

**Fig. 9.** Direct cooling of inverter and motor for compact integration

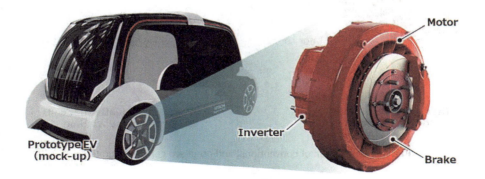

**Fig. 10.** Compact integrated technology for installing drive components inside wheel without remodeling existing suspension

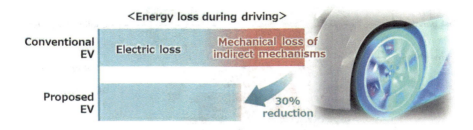

**Fig. 11.** Direct transmission of motor driving force to wheel for energy saving

## 5 Measured Results

A load test and a temperature rise test were performed on the prototype in-wheel system. Figure 12 shows the external view of the test equipment. The in-wheel system was mounted to the test bench, and DC 380 V and cooling oil of 5 L/min were supplied to the inverter.

Figure 13 shows the measured waveforms of voltage and current under the maximum output conditions, which were 60 kW, 600 min$^{-1}$, and 960 Nm. The current waveform is almost a sine wave, although it contains harmonic components due to the carrier frequency of the inverter. It was demonstrated that the prototype could be driven stably as the in-wheel system along with the inverter.

The cooling performance of direct oil cooling was also evaluated. The cooling oil is circulated and cooled by the oil circulation device. The motor current is supplied until the coil temperature is saturated. When the temperature is saturated, the motor current is increased to generate more heat generation, and current supply is continued until the temperature is saturated again. The coil temperature at top and bottom positions was measured with thermistors.

Figure 14 shows the temperature-rise test results. The temperature difference between the upper and lower positions of the coil was about 1 K, and that the temperature-rise trend saturates in about 300 s regardless of the magnitude of the motor current. It was also demonstrated that when the motor current is 50% of the maximum output, the temperature rise of the coil is 45 K. These results indicate that the developed in-wheel system can drive at a sufficiently lower temperature than the allowable temperature of the coil due to the direct cooling.

10  A. Takahashi et al.

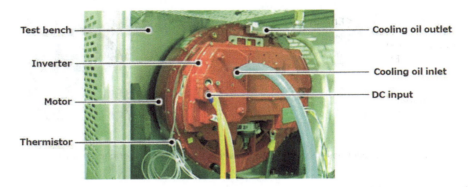

**Fig. 12.** External view of measurement test

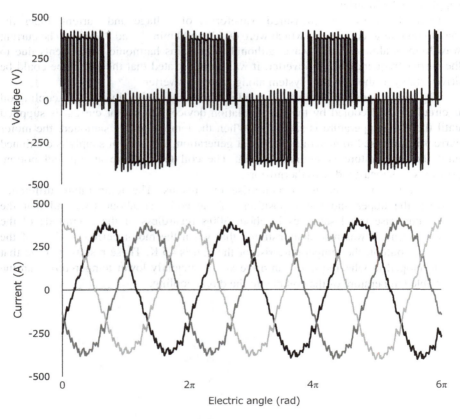

**Fig. 13.** Measured voltage and current waveform (upper: line-line voltage of U and V phases, bottom: U-, V- and W-phase current, DC voltage: 380 V, rotation: 600 min$^{-1}$, torque: 960 Nm)

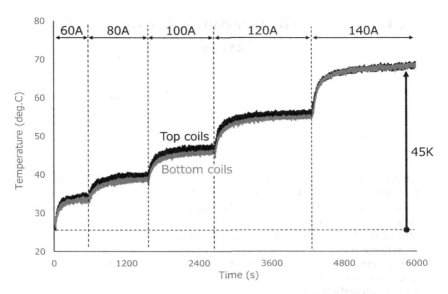

**Fig. 14.** Measured temperature rise of coils (DC voltage: 380 V, rotation: 100 min$^{-1}$)

## 6 Conclusion

We have developed a compact, lightweight direct-drive system for in-wheel EVs. The new motor transmits the high driving force necessary to run an EV directly to the wheels, and its lightweight design minimize the weight increase typically associated with in-wheel units. Furthermore, implementing an in-wheel unit does not require substantial changes to the existing configuration of the suspension and other components.

## References

1. International Energy Agency: Global EV Outlook 2021 (2021)
2. Shao, L., Karci, A.E.H., Tavernini, D., Sorniotti, A., Cheng, M.: Design approaches and control strategies for energy-efficient electric machines for electric vehicles – a review. IEEE Access **8**, 116900–116913 (2020)
3. Sarlioglu, B., Morris, C.T., Han, D., Li, S.: Driving toward accessibility: a review of technological improvements for electric machines, power electronics, and batteries for electric and hybrid vehicles. IEEE Ind. App. Mag. **23**(1), 14–25 (2017)
4. Wang, W., Chen, X., Wang, J.: Motor/generator applications in electrified vehicle chassis – a survey. IEEE Trans Trans Elect **5**(3), 584–601 (2019)
5. https://www.hitachi.com/New/cnews/month/2021/09/210930.html

# $CO_2$- Life Cycle Assessment for the Porsche Taycan

Otmar Bitsche$^{(\boxtimes)}$ and Benjamin Passenberg

Porsche AG, Weissach, Deutschland
`{otmar.bitsche,benjamin.passenberg}@porsche.de`

**Abstract.** The development of the battery electric vehicle is motivated by the reduction of the $CO_2$ emissions. Therefore, it is important to understand the main $CO_2$ contributors during the life cycle of the vehicle. The Life Cycle Assessment of the Taycan confirmed that the battery is responsible for about 40% of the $CO_2$ emissions during the production of the vehicle. This insight leads to the question of the optimal battery size – to meet the customer's needs and the environmental targets. By analyzing customers preferences and use cases, Porsche reinforced following presumptions as an important indication for rightsizing the battery.:

- The majority of daily driving distance is less than 80 km.
- About 80% of weekly driving distance is less than 450 km.
- For long distance travel the customer focuses on traveling time (driving and charging).
- Driving Dynamics is an important criterion for Porsche's customers.
- Green charging energy and optimization of supply chain are main $CO_2$ reduction potentials today
- New cell technologies, higher charging power and increasing recycling ratios are $CO_2$ reduction potentials of the future

In the presentation, we describe how $CO_2$ emissions can be reduced while considering customer needs. Beside using green energy ($CO_2$-neutral) for production and in use charging the battery size/capacity is an important factor affecting $CO_2$ and other environmental factors.

Considering the main factors "$CO_2$", "traveling time" and "driving dynamics" a limitation of battery capacity in combination with fast charging is possible. In order to achieve our Vision of "Net Zero $CO_2$ by 2030" of course recycling is a key success factor.

## 1 Maximize Range: Right Way?

According to a Volvo US study [1] range anxiety is the top barrier for purchasing an EV. Battery size and efficiency of battery electric vehicles was increasing over the last decade. We are seeing long range vehicles like the Porsche Taycan achieving a range of 460 km with one battery charge. Recently announced upcoming releases of battery

---

© Der/die Autor(en), exklusiv lizenziert an Springer Fachmedien Wiesbaden GmbH, ein Teil von Springer Nature 2023
A. Heintzel (Hrsg.): ATZLive 2022, Proceedings, S. 12–20, 2023.
https://doi.org/10.1007/978-3-658-41435-1_2

electric vehicles ranges of 560 and 660 km, touching the upper scale of the range game (Fig. 1). The smaller vehicles used for short trips also a lower range is usually accepted by the customer.

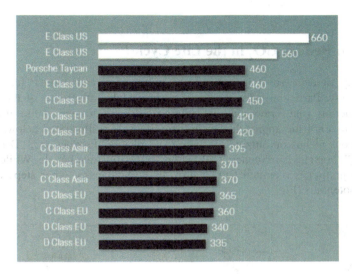

**Fig. 1.** Market perspective on electric range (WLTP) [2]

For seizing the right battery, range is only one trade off factor regarding battery capacity. As shown below, Porsche's customers are traditionally expecting high driving dynamics. In terms of sustainability the commitment to net-zero $CO_2$ is an additional factor to be considered (Fig. 2).

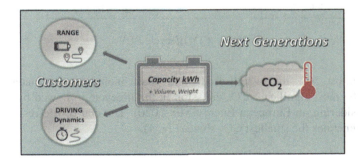

**Fig. 2.** Conflicting factors for "Battery Capacity"

Porsche is committed to $CO_2$ neutrality. In 2030, the company is to be balance sheet $CO_2$ neutral across the entire value chain [3]. The challenge is to fulfill customer demands in all dimensions. Range, driving performance, energy consumption and $CO_2$ footprint. Considering fast charging capability a optimal traveling time can be achieved with smaller batteries.

## 2 How to Reduce $CO_2$ in the Life Cycle?

On the way to net-zero it is not enough to measure and count "Tank-to-Weel" emissions only. Particularly for battery electric vehicles all phases of the life-cycle from production (incl. raw material mining), driving to end-of-life need to be considered in a holistic approach. The holistic lifecycle is illustrated below. Reducing $CO_2$ emissions means optimizing all three main phases "Production", "Driving" and "Recycling". Porsche is influencing all factors in the illustrated cycle. While Taycan's vehicle integration is operated $CO_2$ neutral in Zuffenhausen, other steps offer $CO_2$ reduction potentials (Fig. 3).

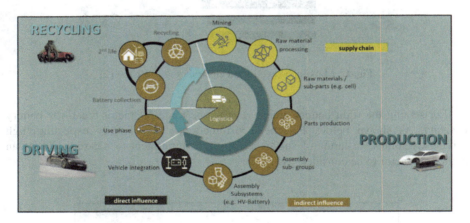

**Fig. 3.** $CO_2$ lifecycle BEV

With almost 50% the overall-production phase including raw material mining generates the most significant $CO_2$ emission in the lifecycle, followed by the driving phase (considering an European Energy Mix 2020). The end-of-life respectively the recycling generates the smallest $CO_2$ share (Fig. 4).

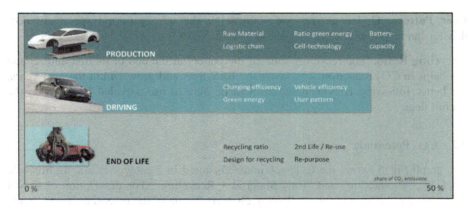

**Fig. 4.** Initial situation of $CO_2$ emissions

## 2.1 $CO_2$ Reduction Potentials

The various sub sections show different $CO_2$ reduction potential. The main influencing areas are described in this chapter.

**Raw Material.** The right selection of the raw material sourcing shows high potential for $CO_2$ reduction. Nevertheless, partnering with the suppliers need to be intensified in the future also in the aspect of the corporate social responsibility (CSR). One of the most effective way to reduce $CO_2$-foodprint of getting raw materials is to reduce the amount of required new raw materials. This is going to be realized by working on progressive recycling concepts, rightsizing of battery capacity, and further optimizing cell technology.

**Ratio Green Energy in Production Chain.** Due to the high energy use in production of e.g. battery cells, steel ... a shift to or an increase of green electricity is essential. Using 100% green energy is essential to achieve Net-Zero.

**Cell-Technology.** Reduction of needed raw materials and new cell technologies reduce $CO_2$.

**Battery Capacity.** The selection of the right battery size and capacity is one of the key factors for $CO_2$.

**Charging Efficiency, Fast Charging and Infrastructure.** The charging losses influence the overall efficiency in the use phase.

Accelerating charging speed in combination with suitable infrastructure will help to eliminate any range anxiety. This will be an enabler for limiting battery capacity.

**User Pattern.** Understanding and influencing the user pattern respectively eco-driving can help to reduce energy use and therefore $CO_2$.

**Recycling Ratio.** The recycling ratio of the battery at the end of its first and second life helps in $CO_2$ reduction. The faster the BEV-fleet will grow within the next years, the better following BEV generations will be able to use recycled Batteries in the overall fleet.

## 2.2 CO$_2$ Potentials

The graph Fig. 5 shows the $CO_2$ potentials clustered in 3 categories among lifecycle "In Use", "Production" and "Raw Material + Recycling". The description of shown potentials can be found in the following chapters. Some of the potentials have already been achieved by Porsche.

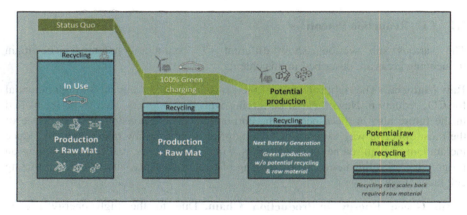

**Fig. 5.** $CO_2$ potentials to Baseline Taycan

**Green Charging Energy.** The final target of 100% green energy is reaching close to zero $CO_2$. For the current production of renewable energy systems like wind turbines (onshore) producing 14 $gCO_2$ per kWh and solar cells (utility) 41 $gCO_2$/kWh [4] needs to be considered as remaining $CO_2$ output. The storage of the intermittent renewable energy needs to be solved in the mid-term. The usage of electric vehicles as grid stabilization and energy buffer can positively contribute to the topic.

**Potentials in the Production.** New production methods particularly for batteries and the use of 100% green energy will further reduce the $CO_2$ footprint. Leveraging the potentials include a strong cooperation with the supplier and raw material value chain.

**Potentials in Raw Materials and Recycling.** This final step in leveraging the potentials can only be achieved by changing the lifecycle illustrated below (Fig. 6).

Most of the raw materials can be replaced by recycled materials in the future. A recycling rate of about 90% is a realistic target on battery level. To get a similar

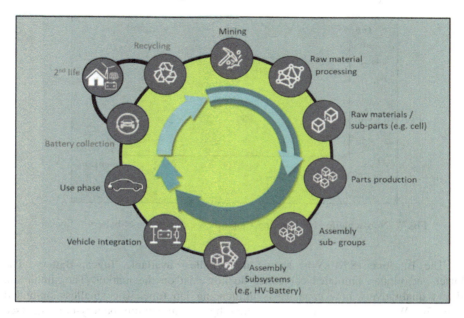

**Fig. 6.** Recycling as a game-changer in lifecycle

recycling rate on the fleet level a certain market penetration of end-of-life BEV is required to "feed the new fleet".

## 3 The Right Sized Battery Capacity?

As one key contributor to the $CO_2$ emissions in the lifecycle it is essential to engineer the right size of the high voltage battery. As showed in $CO_2$-potentials to a certain extend emissions cannot be avoided by 100%. Though battery size is and will be an influencing factor on remaining $CO_2$ emissions.

### 3.1 Conflicting Factor – Driving Dynamics

The expected driving dynamics are often used as the argument for larger battery sizes in the car. Nevertheless, simulating lap times at the Nürburgring are showing different results. The simulation results showing in the figure below considers mass and maximum battery-currents using the same technology (Fig. 7).

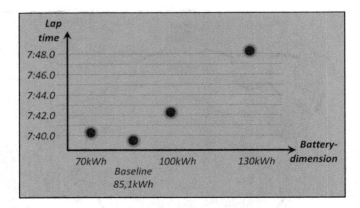

**Fig. 7.** Nordschleife / Nürburgring – lap time vs. dimension of battery – Taycan

The Baseline (85.1 kWh) shows the smallest available Taycan-Battery. The lower and higher simulated capacities do not exist on the market. The simulation shows qualitatively how capacity is influencing lap time. The smallest simulated size (70 kWh usable capacity) is limited by the maximum battery-currents (see 0–200 kph time and lap time). The reduced overall weight (−109 kg or −4.5%) does not compensate the reduced battery size and therefore the reduced battery-capabilities (−17.7%) compared to the baseline variant.

For larger batteries with 100 and 130 kWh useable battery capacity the higher overall vehicle weight creates slower lap times.

### 3.2 Conflicting Factor – Range

In most European countries people do on average 3 trips per day with their passenger car and travel between 30 and 40 km per day [5]. Also in other regions of the world like US and China the majority of the people travel short distances below 100 km per day. The specific driving of the Porsche Taycan customers show a similar picture. 90% of the sum of all daily trips are in the range of 1–120 km. One re-charge is needed at most after 3 days of driving.

### 3.3 Conflicting Factor Charging Time

In the future the charging respectively the charging power itself needs to be optimized according to the charging purpose. For home charging as well as daily charging 11 kW and 22 kW are the preferred choice. Especially 22 kW AC show the highest number of installed charging at the lowest costs. For the 22 kW onboard charger a charging efficiency over 90% can be realized. For long distance travel 150 kW DC and 250 kW DC charging with short charging times are preferred.

### 3.4 Fast Charging as Enabler for Shorter Travel Time

"Remember that time is money" [6] isn't a new finding but time as our most pressures commodity is still valid today. Therefore when looking at long distance travels shortening the travel time needs to get in the focus. Travel time is optimized by the combination of the right battery size and high efficient DC fast charging. The Taycan gains 100 km every five minutes during fast charge. Most of the sources refer to a recommendation of a two hour driving and 15 min break cycle for long distances. Within these 15 min, the Porsche Taycan battery can get enough energy for another two hour drive. Applying this travel behaviour will make travel more save and end discussion about the need for more range.

## 4 What Have We Achieved ($CO_2$)?

### 4.1 Production

Until 2021 certain measures led to a $CO_2$ reduction of 26% between the first and second generation (not yet on the market) of battery electric vehicle. The measures are design optimization, material reduction and supply chain optimization.

### 4.2 In-Use

Most of the Porsche customers are using green charging solutions at home or at the public charging point. In addition, Porsche actively works on the reduction in the losses due to auxiliary loads and achieved already highest efficiency with the 22 kW on-board charger. Already since beginning 2020 Porsche offers with selected charging partner access to green fast charging networks [7].

## 5 What's Next?

In the mid-term and long-term future, the target is a balance of usability and further $CO_2$ reduction.

### 5.1 Focus Area Recycling

Massive recycling investments will create a change in lifecycle. The high market penetration of battery electric vehicles and HV batteries will generate scalable recycling benefits. The needed technology advances can be only achieved via strong partnerships in the whole supply chain.

### 5.2 Focus Area Technology

New cell technologies will reduce the energy use. Higher charging power will increase the efficiency. Advanced cooling concepts and thermo management complete the picture. With these two focus areas the successors of the current Porsche Taycan will achieve $CO_2$ neutrality in a circular economy in the future.

# References

1. Volvo Car US: Poll Finds Americans Feel Electric Vehicles Are The Future of Driving. https://www.media.volvocars.com/us/en-us/media/pressreleases/248305/poll-finds-americans-feel-electric-vehicles-are-the-future-of-driving. Accessed: 23. Aug. 2021
2. EV Database: https://ev-database.org
3. PAG: 18/03/2021, Porsche aims for balance sheet $CO_2$ neutrality in 2030. https://www.porsche.com/international/aboutporsche/e-performance/magazine/co2-neutrality-2030/. Accessed: 23, Aug. 2021
4. Joshua, D.R.: Nuclear and wind power estimated to have lowest levelized $CO_2$ emissions. https://energy.utexas.edu/news/nuclear-and-wind-power-estimated-have-lowest-levelized-co2-emissions. Accessed: 24. Aug. 2021 (2017)
5. Eurostat: https://ec.europa.eu/eurostat/documents/3433488/5298257/KS-SF-07-087-EN.PDF.pdf/bf69235f-f285-4dc0-ac55-55a2d0c69c11?t=1414687808000. Accessed: 24. Aug. 2021 (2007)
6. Franklin, B.: Advice to a Young Tradesman (21 July 1748)
7. PAG: https://newsroom.porsche.com/en/2020/company/porsche-charging-service-destination-charging-ionity-taycan-delivery-start-europe-19824.html. Accessed: 24. Aug. 2021

# ZF eConnect: Efficient Solutions for AWD BEV

Alessio Paone[1]([✉]), Stephan Demmerer[2], Matthias Winkel[2],
Martin Ruider[2], Philip Endres[2], and Uwe Großgebauer[2]

[1] ZF Friedrichshafen AG, Schweinfurt, Germany
alessio.paone@zf.com
[2] ZF Friedrichshafen AG, Friedrichshafen, Germany
{stephan.demmerer,matthias.winkel,
martin.ruider,philip.endres,uwe.grossgebauer}@zf.com

**Abstract.** Among the battery electric vehicles (BEV) a trend towards all-wheel drive (AWD) is visible and the AWD market share is going to grow significantly within the next years. For this reason, it is important to find a way how to increase the overall powertrain efficiency. One possible solution is to disconnect the secondary axle every time the driver demand doesn't require it to be active. We call this solution eConnect. In this work, different eConnect solutions for AWD BEV and their advantages in terms of efficiency are discussed. In the most AWD BEV (with two electrified axles) the primary axle is used to cover most of the common driving situations. The secondary axle (also boost axle) is activated only for a low percentage of the vehicle lifetime in order to achieve better acceleration or to fulfill particular use cases. It is proven that disconnecting the secondary axle reduces the drag torques and significantly increases the powertrain overall efficiency, increasing the vehicle electric range. This is true also for secondary axles with ASM. ZF is working on different eConnect solution in order to fulfill the customer requirements in terms of performance and costs. One of these solutions was already implemented in a vehicle as a prototype with good results.

**Keywords:** Components for electrified powertrain · eConnect · Efficiency

## 1 Overview of Losses in BEV AWD Vehicles and Why it is Worth it to Disconnect the Secondary Axle

In an electric drive for passenger cars conventionally made up of an inverter, an electric motor and a reducer gearset, several different kinds of electrical and mechanical losses occur, since each component cannot operate with a 100% efficiency.

---

© Der/die Autor(en), exklusiv lizenziert an Springer Fachmedien Wiesbaden GmbH,
ein Teil von Springer Nature 2023
A. Heintzel (Hrsg.): ATZLive 2022, Proceedings, S. 21–32, 2023.
https://doi.org/10.1007/978-3-658-41435-1_3

Discussing and analyzing in detail all the losses in an eVD is not the scope of this chapter, nevertheless it is important to understand the sources of the losses in order to find feasible ways to reduce them, since reducing the losses means increasing the overall efficiency of the vehicle hence the electric range.

First of all, we can divide the losses into two categories, which we will call dynamic and static losses. The dynamic losses are related to the masses and the inertias that have to be moved in the vehicle and also to the drive strategy and driver's behavior. The static losses are related to the electrical and mechanical components and are the consequence of transformation of the electrical into mechanical power and the torque multiplication and speed reduction (Fig. 1).

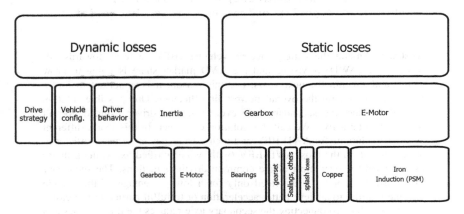

**Fig. 1.** Losses within an eVD system

Considering an average eVD system which is currently under development at ZF and based on the WLTP we can assess that the losses in the inverter are approximatively 10% of the overall losses, whereas the electrical losses in the e-motor are ca. 60%. The rest, 30%, are mechanical losses in the reducer gearset.

The electrical losses coming from the e-motor are basically iron, copper, magnet losses. Some mechanical losses can also be found in the e-motor because of the air resistance, some splash losses depending on the cooling system (e.g., oil), losses coming from bearings and seals. The losses related to the e-motor vary strongly depending on the type on motor (e.g., ASM, PSM, SESM) (Fig. 2).

ZF eConnect: Efficient Solutions for AWD BEV 23

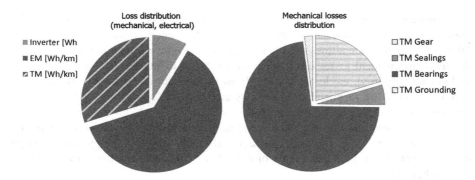

**Fig. 2.** Loss distribution within an eVD system

The mechanical losses in the reducer are related to the friction between the gears, the seals, the bearing and also to the lubrication on the gearset itself. The mechanical losses also depend on the particular design and the topology of the gearset. We can mainly consider single-stage or two-stage reducer and generally assume that the former has better characteristics in terms of losses than the latter.

Based on the WLTP, these losses can be calculated, simulated or measured, depending on the development stage. An average simulation of the losses in shown in Fig. 3. We can see that the electrical losses are usually bigger than the mechanical ones, and the total losses can vary within a range of 20 to 25 Wh/km depending on the design of the e-motor and the reducer.

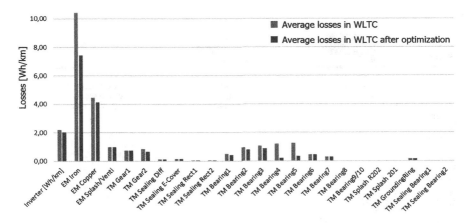

**Fig. 3.** Losses calculated based on the WLTC

Considering that, and considering a BEV AWD passenger car, it is theoretically always worth it to disconnect the secondary axle every time it is not in use. Disconnecting the axle allows to save the dynamical losses of the rotating inertias and the static, mechanical losses of the gearbox. In the moments the axle is only dragged and not actively used, the losses coming from the e-motor are quite low. If the motor is a PSM type, the magnetic drag torques can be quite significant and the overall savings through a disconnect unit increase.

The loss saving and the efficiency increase have always to be considered together with the additional costs that a disconnect unit causes and the energy costs coming from the battery. At the current state of the art, we think that the brake even point is not reached yet.

## 2 Efficiency Increase Based on a Comparison with a Fastback Mid-Size Four-Door Sedan

In order to determine if and how much a disconnect unit is worth to consider for a passenger car, ZF has made several investigations considering one of the best fastback mid-size four-door sedan in terms of mechanical losses available on the market. These investigations are based on simulations of the WLTC with a 75 kWh battery. The secondary axle has an ASM motor and a very efficient two-stage reducer. These are the hardest condition to test if a disconnect unit is able to significantly increase the overall vehicle efficiency, since the ASM motor has no magnetic drag torques.

The results of the simulations are summarized in Fig. 4.

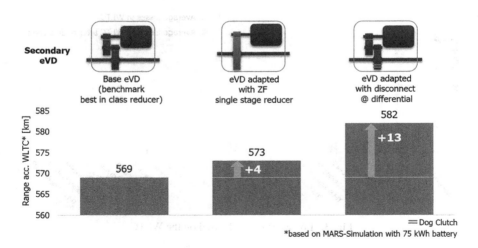

**Fig. 4.** Advantages of disconnect system in terms of electrical range based on WLTC

Frist of all we have simulated the maximum electrical range in the WLTC with the current secondary axle configuration. The simulation is based on the assumption that during the WLTC the secondary axle is used only occasionally, since the cycle doesn't require it in terms of performance. This of course depends on the drive strategy of the vehicle, but we think that assuming that the secondary axle is active for only (approximatively) 5% of the cycle is a good estimation.

On this basis, we have calculated the maximum electrical range.

In the second step, we have performed the same calculation by substituting the current two-stage reducer with a single-stage one, which is currently the best possible reducer system in terms of mechanical losses. This brings an increase of 0.88% on the electrical range.

The third calculation shows the effect of introducing a disconnect unit in the differential gear, which brings an increase of 2.28%. This is quite significant and shows how, even in a very efficient eVD system, a disconnect unit can increase the vehicle efficiency.

We have considered here a disconnect unit located in the differential. There are several options where to disconnect the axle. We will discuss these in the next chapter.

## 3 Possible Topological Layouts

In order to describe the possible topological layouts of a disconnect unit, we will consider a two-stage reducer gearset with the e-motor and the side shafts on different axis (parallel axis layout). This is one of the most common layouts available on the market.

There are basically four feasible options where to design the disconnect unit (Fig. 5):

- on the side shaft (position A)
- in the differential gear (position B)
- on the intermediate shaft (position C)
- on the rotor shaft (position D)

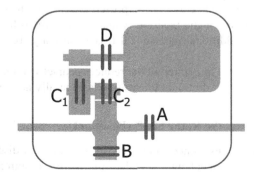

**Fig. 5.** Topological layouts for disconnect system

Each of these design options has advantages as well as disadvantages that we are going to describe more in detail.

## 3.1 Disconnect Unit on the Side Shaft (Position A)

The disconnect unit on the side shaft allows the maximum reduction of power losses because the whole gearset is disconnected and almost all the static and dynamic losses out on the reducer are eliminated. Only residual bearing and possible sealing losses are left over. This solution is quite independent from the eVD design and can be considered as an add-on component which possibly doesn't interfere with other eVD components like gearset, differential, housing.

This solution must cope whit the highest rotational speeds and the biggest inertias, that can make the synchronization process more challenging. Also, the transmitted torques are very high. Due to the position outside the eVD system, a dedicated lubrication of the mechanical components is necessary. Furthermore, it is very difficult to implement this solution for an eVD system with the e-motor on the same axis as the side shafts (coaxial layout) because of the packaging restriction in axial direction.

## 3.2 Disconnect Unit in the Differential Gear (Position B)

If the disconnect unit is integrated into the differential gear, this also leads to a very good reduction of the mechanical losses, comparable with the solution in the position A. This solution has to be fully integrated in the reducer gearset which makes the supply of lubrication much easier. Furthermore, the integration in the differential gear allows to design the disconnect unit optimizing the packaging requirements.

The biggest disadvantage presented by this solution is the highest torque to be transmitted in comparison with the other solutions. Also, the inertias that must be synchronized are quite high.

## 3.3 Disconnect Unit on the Intermediate Shaft (Position C)

The intermediate shaft represents a good compromise between torques, inertias that need to be synchronized and reduction of power losses. The loss reduction for this solution is not as high as for the position A and B since the differential gear is still rotating, which brings additional mechanical losses basically from the gears and the bearings.

But this solution can be integrated in the reducer gearset without or with very little addition installation space. Additional lubrication is generally not needed.

## 3.4 Disconnect Unit on the Rotor Shaft (Position D)

The last position for the disconnect unit that we are going to discuss is the one on the rotor shaft. The disconnect unit here has the lowest requirements in terms of transmitted torques, since it is located ahead of the reducer gearset. This also leads to the lowest possible inertias.

On the other hand, and for the same reason, this disconnect unit has to cope with the highest rotational speeds. This can make the control of the synchronization process very challenging. But the biggest disadvantage of this solution is represented by the fact that the reduction of the mechanical losses can be very low and basically not worth it because the whole gearset is still rotating. Only the mechanical losses caused by the e-motor are disconnect and these are always much lower as the losses caused by the transmission. For this reason the solution D is currently not under development at ZF and we will not consider it anymore.

In the following table, the advantages and disadvantages of the four topological layouts for the disconnect unit are summarized (Table 1):

**Table 1.** Advantages and disadvantages of disconnect layouts

| Layout | installation space | loss reduction | differential speed | torque |
|---|---|---|---|---|
| **A** - side shaft | 0 | ++ | + | − |
| **B** - differential | + | ++ | ++ | − |
| **C** - intermediate shaft | ++ | + | + | 0 |
| **D** - rotor shaft | + | − | − − | + |

# 4  ZF Design Solutions

In the last period ZF has worked on several solutions for a disconnect unit which can fulfill the market and customer requirements. Different technical concepts were analyzed and compared in order to find the best possible solution in terms of technology and costs.

In the following paragraphs we will take a closer look on two of these solutions.

## 4.1  eConnect – ZF Modular Kit

In order to find the best solution for its customers, ZF has put the focus both on the mechanical and the electrical/electronic side of the disconnect technology.

On the mechanical side, we have chosen a dog clutch as a form fit decoupling element. The dog clutch allows to reduce the shifting forces and actuation time to a minimum and can be controlled to minimize the noise and the shifting jerks. Other solution like friction clutches were also analyzed but those bring disadvantages in terms of shifting energy, packaging and costs.

The dog clutch is actuated by a DC servomotor. The electromechanical actuation is chosen because of its flexibility, low packaging demand, low shifting time and also low costs. Other solutions like hydraulic actuation weren't taken into consideration because of many disadvantages in the categories mentioned above.

Another common possibility for the actuation device is based on the electromechanical technology with a solenoid actuator. This solution can be competitive in terms of costs, but a deep investigation showed that the electromechanical solution has the best results particularly for the shifting quality and the shifting energy, so ZF decided to build its modular eConnect kit based on that technology.

The ZF DC servomotor is bistable, so that no current is needed neither in the open nor in the closed position.

This approach allows to develop a very flexible modular kit which can be easily adapted to the layout described in the previous chapter.

To summarize, the ZF modular kit eConnect has the following characteristics:

- Dog clutch with bi-stable actuator system
- Actuation with DC servomotor and clutch position info (ASIL B from C)
- Complete systems approach for optimized acoustic behavior
- Adaption of shifting process to current driving situation based on vehicle command
- Lifetime lubrication filling for side shaft without exchange between DCU and reducer possible

The key parameters of this product are summarized in the Table 2.

**Table 2.** Key parameters of ZF eConnect solution

| | |
|---|---|
| Shifting cycles | 600.000 |
| Power supply [V] Operational \| full-dynamic | 9–10.6 \| 10.6–16 |
| Max. peak current [A] | 20 |
| Ambient temperature [°C] Operational \| full-dynamic | −40 up to 120 \| −20 up to 100 |
| Control | Integrated/External |
| Communication | CAN FD |
| Functional safety | ASIL B (C) |
| Safe state | Connected/disconnected |
| Shift element | Dog clutch (form-fit) |
| Power (end position) | <2W (bi-stable) |
| Connecting time @50rpm [ms] | <100 |
| Disconnecting time @load free [ms] | <100 |
| Max. disconnect torque @side shaft [Nm] | 15 |

The ZF eConnect is also fully integrated in the low voltage subcomponents architecture, as showed in Fig. 6:

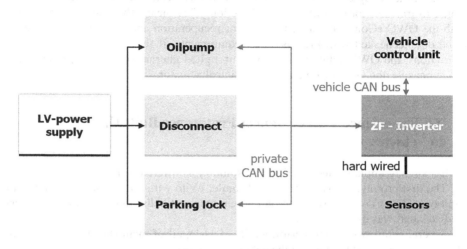

**Fig. 6.** Low voltage slave architecture of ZF eConnect

The main characteristics of this integration are:

- High flexibility regarding subcomponent integration
- Not dependent on customer interface, enabler for multi-customer approach
- Less effort for integration to inverter
- Simplification of requirements regarding cybersecurity
- Reduced bus load for vehicle CAN
- Cost-effective voltage supply for subcomponents (no inverter integration, reduced EMC efforts)

### 4.2 eConnect with OWC (One-Way-Clutch)

One particular eConnect solution ZF is developing for the low-cost low-segment market is based on a one-way-clutch.

This is a very simple solution based solely on the mechanical component OWC, which doesn't require any shifting control and electronic components. Only some adjustments on the e-motor speed and torque control are needed.

ZF is developing this solution for the side shaft (layout A).

Due to the mechanical characteristics of a one-way-clutch, every time the e-motor supplies torque and its speed (reduced on the side shaft) would become higher than the wheel speed, the one-way-clutch engages and transmits this torque.

First simulation results are very positive and show a smooth behavior by engaging and disengaging. A prototype has been built up and will be tested in the vehicle soon.

This solution has the biggest advantages in terms of installation space, complexity and, above all, costs.

The disadvantages are related to the mechanical characteristic of the one-way-clutch to transmit torque only in one direction. This means that the secondary axle with the OWC eConnect cannot be used for recuperation and rear drive. This needs to be doublechecked with the customer requirements. If the customer accepts these restrictions, the OWC solution can represent a good alternative to the more complex and expensive dog clutch with electromechanical actuator.

## 5 ZF Experience with eConnect Prototypes Built Up in Vehicles

ZF has already implemented its eConnect solution in different vehicles.

The first prototype was built for the Daimler eVito with a ZF series front axle. An additional ZF eVD was assembled on the rear axle and a disconnect unit on the intermediate shaft was implemented.

Several simulations were performed in advance, all of them showed the potential of the disconnect unit to increase the vehicle efficiency.

Measurements with the prototype vehicle based on the WLTC were finally performed and these confirmed the simulation results.

The mechanical losses were reduced up to 90% and the reduction in the WLTC was 1.4 kWh/100 km, which is significant (Fig. 7).

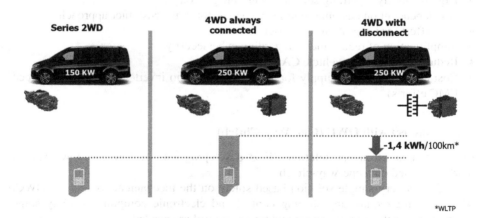

**Fig. 7.** ZF prototype experience with Daimler eVito

Another prototype has been built up in the last months for an application in the fastback mid-size four-door sedan already mentioned in Chap. 2. In this prototype the eConnect is placed on the side shaft. The e-motor is an ASM type and the original reducer is considered to be one of the best reducers currently available on the market in terms of efficiency.

On the test bench the original axle, a ZF axle currently under development and the prototype axle with eConnect were measured and compared, as showed in Fig. 8.

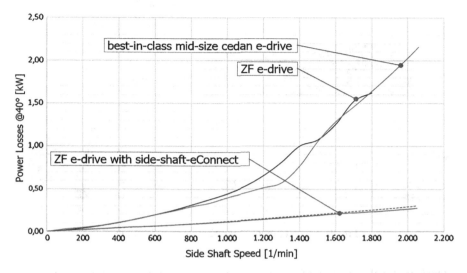

**Fig. 8.** Losses comparison between best-in-class e-drive and ZF e-drive with eConnect

The measurements confirm all the investigations and simulations done before. The ZF eConnect system shows a significant reduction of the losses and contributes to increase the overall vehicle efficiency and the electrical range.

## 6 Market Potentials

If we take a look to the development of the market in the following years, we definitively see a rapid growth in the sales of vehicles with eVD applications (BEV, HEV and PHEV).

For possible eConnect applications, the AWD market penetration is important. Based on ZF estimations, a stable penetration of approximately 30% can be forecasted. We estimate a growth for the BEV AWD market up to 14 million vehicles until 2030. While the entire market for secondary axle can in principle be considered, implementing disconnect solutions might be limited to applications where the secondary axles are based on PSM technology. On the other side, we have also proved (by simulations and measurements) that eConnect allows a significant reduction of the losses even with the ASM technology for the secondary axle (Fig. 9).

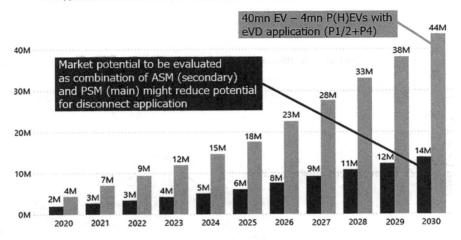

**Fig. 9.** Market potential for eConnect applications

Decisive for the success of eConnect are on one side the energy costs and on the other side the costs of eConnect itself. Based on our current analysis of the market, the feedback gathered from different customers on different markets (NAFTA, Asia) and the latest cost estimations of the ZF eConnect technology, we think that the product has very good chances to be established on the BEV AWD market within the following years.

# Balancing of Efficiency, Costs and $CO_2$-Footprint for Future Mobility

Christoph Danzer[1]($\boxtimes$), Alexander Poppitz[1], Tobias Voigt[1], Manfred Prüger[1], and Marc Sens[2]

[1] IAV GmbH, Stollberg, Germany
```
{christoph.danzer,alexander.poppitz,
        tobias.voigt}@iav.de
```
[2] IAV GmbH, Berlin, Germany
```
marc.sens@iav.de
```

**Abstract.** The future powertrain mix will not only be influenced by technology and engineering. The properties of fuels, the production of electric energy and the footprint of all mobility elements are mandatory to achieve the global climate targets. To address this, IAV connected it's technology oriented powertrain simulation methods with the energy and fuel production assessment in one method-chain. In application of this method-chain IAV has investigated powertrain systems with fossil, hydrogen and synthetic fuels and compared it with electricity-based mobility. The results of this study will be presented in order to get an overview about the powertrain effects one the one hand and the cost and $CO_2$-impact of the energy and fuel production. Beside this the complete life cycle assessment of vehicle with powertrains and the energy/fuel provision will be presented for different detail levels. Based on that results IAV will give a big picture and a clear recommendation about the best balanced powertrain systems in regard of consumption, powertrain footprint, energy/fuel footprint, costs and TCO. Beside this the impact of those systems on the future $CO_2$-fleet-emissions and the ecological potentials will be addressed.

**Keywords:** Life Cycle Assessment · Future Mobility · Powertrain · Hydrogen

## 1 Introduction

The objective of the comparative study is the universal evaluation of hydrogen-based powertrain systems, including hydrogen production, and their comparison with powertrains using fossil fuels and battery-electric powertrains. The Well-to-Wheel (WtW) consideration is additionally extended by the analysis of the complete vehicle life cycle (LCA), which enables the environmental impact of the production and recycling of the vehicles and powertrains from Cradle to Grave (CtG). Furthermore, the economic aspects of hydrogen production and propulsion system deployment

---

© Der/die Autor(en), exklusiv lizenziert an Springer Fachmedien Wiesbaden GmbH, ein Teil von Springer Nature 2023
A. Heintzel (Hrsg.): ATZLive 2022, Proceedings, S. 33–45, 2023.
https://doi.org/10.1007/978-3-658-41435-1_4

are considered, leading to the specification of manufacturing costs and total cost of ownership (TCO)

All powertrain systems in all vehicle classes were individually optimized in their main parameters and examined under the same operating boundary conditions. This is based on IAV's unique Powertrain Synthesis methodology [1, 2] with coupled life cycle and cost assessment. The results make it possible to compare different powertrain concepts in different vehicle classes, taking into account the production paths of hydrogen, electrical energy, the powertrains and the vehicles with regard to technical, ecological and economic aspects

## 2 Techno-Economical Study of Powertrain Systems and H$_2$-Production Paths

Figure 1 schematically shows the considered H$_2$ generation paths, the vehicle segments and the powertrain types. For the production of hydrogen, the processes steam methane reforming (SMR) – grey hydrogen, SMR with carbon capture and storage (CCS) – blue hydrogen, methane pyrolysis – turquoise hydrogen and water electrolysis with different electricity mixes are considered. In addition, the import of renewable hydrogen from MENA (Middle East North Africa) countries is considered. In addition to the feedstock's, the required electrical energy is considered in a differentiated manner, either from the electricity mix anticipated for 2030 [3] or exclusively from renewable generation.

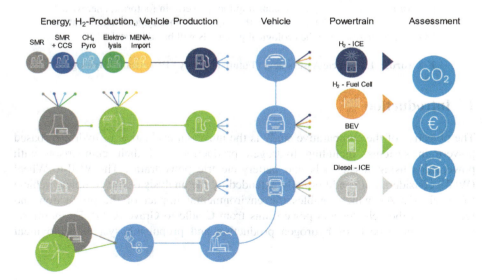

**Fig. 1.** Overview of H$_2$-Production paths, energy supply, vehicle segments and powertrain types

This differentiation is also applied to the charging processes of the BEV's and partly to the production of the vehicles and powertrain components.

On the vehicle side, this study differentiates in passenger cars (SUV-segment) and commercial vehicles (light and heavy). Each vehicle segment is considered with $H_2$ combustion engine powertrain ($H_2$-ICE), $H_2$ fuel cell powertrain ($H_2$-FCEV) and pure battery powertrain (BEV). In addition, diesel combustion engines (Diesel-ICE) are used as fossil reference systems for evaluation. The most important boundaries of the study are illustrated in Table 1.

**Table 1.** Vehicle and Study Boundaries

| Vehicle | Passenger Car | Light Commercial Vehicle | Heavy Commercial Vehicle |
|---|---|---|---|
| Basis simulation weight [kg] | 1600 | 2400 | 35,000 |
| Powered axles [-] | FWD | RWD | RWD |
| Range [km] | 500 | 500 | 800 |
| Cycle [-] | WLTP | WLTP | VECTO Long-Haul |
| Mileage [km] | 200,000 | 200,000 | 1000,000 |
| General electricity mix [g $CO_2$e / kWh] | 220 (fossil and renewable grid mix) | | |
| Renewable electricity mix [g $CO_2$e / kWh] | 24 (25% PV / 75% Wind) | | |
| Target year [-] | 2030 | | |

# 3  Results for Passenger Cars

For the passenger car segment, a compact class SUV with a range of 500 km was assumed. For the 90 kW $H_2$-ICE, a transmission optimization for a 6-speed dual clutch transmission was performed. The fuel cell propulsion system was optimized in terms of HV battery energy capacity to 4.25 kWh and key axle drive parameters (EM power, EM torque, number of speeds, and transmission ratios). The axle drive parameters were also varied for the BEV. Both hydrogen-powered systems, as well as the battery electric powertrain, were designed to always have the same target range. The number of powertrain concepts considered in the passenger car thereby comprises approximately 68,300 variants.

The FCEV system with an optimized operating strategy achieves a consumption advantage in the WLTP of approx. 40% compared to the optimal $H_2$-ICE system. Here, the fuel cell is often operated in low power operating points with highest efficiencies. The additional consumption potential due to hybridization for the ICE-variant is about 19% for HEV technology. Based on the results of the powertrain optimization and the $H_2$ as well as electricity production paths considered, the respective $CO_2$ emissions (WtW) were determined and compared with those of the battery- and diesel-powered vehicles.

For a hydrogen production in Germany with 2030 electricity mix, blue hydrogen represents the lowest $CO_2$ pathway according to well-to-wheel balances, as shown in Fig. 2. With low-$CO_2$ hydrogen from the MENA region, $CO_2$ emissions in both hydrogen-powertrains systems could even be reduced to about half compared to blue hydrogen. With blue hydrogen produced with 2030 electricity mix, a fuel cell vehicle would then be about 10 g $CO_2$e/km better than a battery vehicle charged with the same electricity mix and 33 g/km worse if it would be possible to charge the BEV with renewably generated electricity. Only if 100% renewable electricity can be used for hydrogen production, electrolysis represents a considerable $CO_2$ potential. Looking at the entire vehicle life cycle both hydrogen powertrains reach almost the same $CO_2$ level as the battery vehicle, even if this is charged with 100% renewable energy. It is highly likely that hydrogen can be produced in larger quantities with renewable energy in 2030. Assuming this is the case, the considered hydrogen powertrains would even emit about 40 g $CO_2$e per kilometer less than a battery vehicle, which is charged in average with ordinary electricity mix. In total, the comparison shows that in case of using renewable electricity for the passenger cars segment, the $CO_2$ backpack of the battery vehicles almost balances out with the additional emissions of $H_2$ generation over the vehicle life cycle.

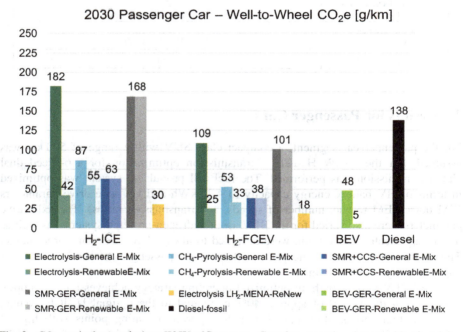

**Fig. 2.** $CO_2$ equivalent emissions WtW of Passenger Cars for general and renewable electricity mix in 2030

Looking on TCO and production costs, an annual number of 500,000 units per year was assumed for all powertrain systems in 2030. Despite the larger storage volume of the $H_2$-ICE system due to consumption, the component costs of the FCEV systems are around 1000 € higher. Due to the tank technology, both hydrogen powertrains are in total about 2000 to 3000 € more expensive than the diesel reference. Looking at the TCO values for 2030, it is noticeable that when using MENA-hydrogen, the total costs are at the same level as diesel and even 3 to 4 €ct/km cheaper than with a battery vehicle. In terms of TCO, no clear trend can be identified for either of the two hydrogen propulsion systems.

The well-to-wheel $CO_2$ potential of the examined passenger cars with hydrogen powertrains is significant with 54% compared to a corresponding diesel vehicle, even if the hydrogen is produced in 2030 with ordinary electricity mix. In the considered passenger car SUV segment, both the $H_2$-ICE and the FCEV powertrains with blue hydrogen are on a similar level in terms of $CO_2$ as the battery powertrains. With MENA imported hydrogen and fuel cell propulsion system, $CO_2$-emissions can be reduced to only 18 g $CO_2$e/km (WtW, WLTP). Considering the whole vehicle life cycle, it is clear that with imported MENA- $H_2$, both hydrogen powertrains can reach almost the same $CO_2$ level of a battery vehicle, even if this would be charged with 100% renewable energy. In terms of overall costs, the hydrogen propulsion systems are even cheaper than a comparable battery vehicle. Overall, for the passenger car segment, the fuel cell system shows the greatest potential in terms of efficiency, cost and sustainability.

# 4 Results for Light Commercial Vehicles

The powertrain systems in the light commercial vehicle segment were also optimized in the same way as for passenger cars. From the total concept amount of approx. 192,000 variants, the consumption-optimized systems were selected in relation to a 500 km range. According to the results, the consumption advantage of the fuel cell powertrain system is reduced to about 28% compared to the $H_2$-ICE. This is mainly due to the specific higher utilization of the LCV-ICE compared to the passenger car application. The optimization resulted in a capacity of 9.25 kWh as the most fuel-efficient battery size for the FCEV systems. The consideration of the WtW-$CO_2$ balances in Fig. 3 shows that in the LCV, the combustion engine with blue hydrogen can achieve slightly lower WtW-$CO_2$ emissions than an equivalent battery vehicle with 2030 electricity mix.

Furthermore, it can be seen that with imported MENA hydrogen, the equivalent $CO_2$ emissions are between a factor of two to three lower than with a battery vehicle charged with 2030 electricity mix. A $H_2$-ICE vehicle with blue hydrogen would be approximately 50% lower in $CO_2$ (WtW) than a diesel-powered vehicle.

Under current TtW-legislation, average OEM $CO_2$-fleet emissions are reduced primarily by zero-emission vehicles, with an almost linear increase in manufacturing costs on the fleet level. Expanding the $CO_2$-balancing limits, for example to WtW, would result in an absolute increase in emissions as well as greater potential for optimizing costs. Figure 3 (on the right) also shows a high cost gradient as soon as WtW-emissions are reduced to values below approx. 120 g $CO_2$e/km with the predicted 2030 electricity mix. It is not possible to bring the WtW-value below approx. 80 g $CO_2$eq./km, even with 100% battery electric vehicles. $CO_2$-neutrality can only be achieved by further increasing the share of renewable energy compared to the 2030 electricity mix.

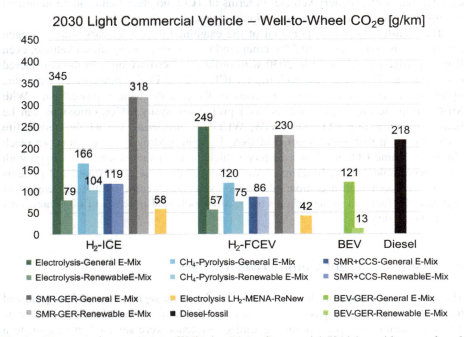

**Fig. 3.** $CO_2$ equivalent emissions WtW for Light Commercial Vehicles with general and renewable electricity mix in 2030

With 100% renewable electricity, the resulting WtW-emissions for the electrolysis and methane pyrolysis pathways can be lowered below the level of blue hydrogen. Assuming an all-seasons target range of 500 km, the BEV systems have a battery energy content of more than 300 kWh, which has a significant impact on the $CO_2$ footprint of the powertrain.

Analyzing the life cycle $CO_2$ emissions of the different powertrains, which include both the electrical energy and the vehicle production. Here, the hydrogen powertrains show a very similar level compared to the BEV variants almost regardless of the generation path. It can be deduced from this that in case of using renewable energy the operating emissions of the various powertrain systems and the $CO_2$ footprints of the components almost balance each other out in terms of the entire life cycle. If the battery vehicle is charged with ordinary electricity mix, blue and turquoise hydrogen result in significant $CO_2$ potentials for the hydrogen powertrain of up to 86 g $CO_2$e/ km assuming the predicted electricity mix for 2030.

The high energy content of the EV battery systems is reflected not only in a high impact on the $CO_2$ footprint, but also in increased production costs. Even assuming full carry over use of passenger car cell modules, the total cost of the powertrain system in the LCV rises to over 30,000 €. The more powerful fuel cell, battery and e-drive unit (EDU) compared to the passenger car increases the production costs of the FC-powertrain by about 3000 € compared to the $H_2$-ICE powertrain, despite the smaller tank system. The powertrain scenarios with ICE propulsion are even cheaper overall than the FC-systems, despite the additional hydrogen consumption. Nevertheless, in terms of cost, diesel is the cheapest propulsion system at 30 €ct/km. Assuming that the high production costs of the battery systems are passed on to the sales prices in the same way as for the other powertrains, the BEV systems represent the most expensive mobility scenario for this vehicle segment at around 48 to 50 €ct/km.

For the light commercial vehicle, there is particular potential for the $H_2$-ICE powertrain. Compared to passenger cars, the combustion engine has a higher specific load, which means that the consumption gap to the FCEV powertrains are smaller in this vehicle segment. This results in a relatively compact and cost-efficient system settings. Both the FC- and ICE-powertrain offer lower WtW-$CO_2$ emissions per kilometer in 2030 using blue hydrogen compared with a BEV using ordinary electricity mix. Another $CO_2$ and cost potential is possible by using renewable MENA hydrogen, where in terms of TCO the $H_2$-ICE is even the cheapest scenario among the low-$CO_2$ systems. Only fossil diesel is still expected to be about 4 €ct/km cheaper in 2030 without any fiscal intervention. Accordingly, the FC-powertrain and the $H_2$-ICE are interesting alternatives to fossil and battery electric powertrains under all evaluation criteria. Based on the premises made, there are slight advantages for the $H_2$-ICE powertrain type in the LCV segment.

## 5  Results for Heavy Commercial Vehicles

Due to the variance of application scenarios, the heavy-duty commercial vehicles segment is characterized by a high degree of diversification in the requirements placed on the powertrain system. In addition, legal requirements, in particular exhaust emission and $CO_2$ legislation, determine the powertrain layout. The $CO_2$ legislation of the European Union (EU), which stipulates a $CO_2$ reduction in the heavy on-high-way sector of 15% by 2025 and 30% by 2030 compared to 2019, is particularly noteworthy here. In addition, entry bans for inner cities with internal combustion engines are to be expected, which means that electric drives will come into focus, especially in the heavy regional delivery sector. In this study, the focus is on the application in heavy-duty long-haul commercial vehicles. Heavy delivery trucks are

40  C. Danzer et al.

discussed comparatively. The resulting optimum powertrain system configurations show that the BEV has the lowest energy consumption, followed by $H_2$-FCEV, Diesel and $H_2$-ICE. However, the percentage improvements in consumption between 2025 and 2030 differ between the powertrain types. For Diesel, an efficiency increase of about 10% is possible with the help of a high-efficiency concept. This includes waste heat recovery, phase change cooling, friction optimization, an optimized injection system and intelligent thermal management. The hydrogen variants also improve significantly towards 2030, the FCEV benefiting more than the $H_2$-ICE truck, thanks to the application of an intelligent and predictive operating strategy and the resulting higher recuperation potential.

In addition to efficiency, technological maturity is crucial for market penetration. It is expected that the $H_2$-ICE can go into series production as early as 2024, thus offering a short term $CO_2$ reduction potential. Fuel cell technology currently does not yet meet all the robustness requirements of a long-haul application and must be operated with high-purity hydrogen. In the long term, however, the $H_2$-FCEV offers the optimal long-haul propulsion system. A highly efficient Diesel is attractive to leverage $CO_2$ potential in the short term in combination with zero emission vehicles in the fleet, and only offers full potential in the long term if e-fuels are approved as zero emission fuels by future legislation. A long-haul BEV should be seen as a solution for selected and appropriate use cases. Even with further development of the battery technology, the geometric integration of the cell modules, as well as charging infrastructure and reduction of the payload remain as challenges.

The Heavy Regional Delivery sector presents a differentiated picture. As described at the beginning, however, the application profiles of the applications and the needs of the fleet operators differ. Due to political and public pressure, the BEV is coming to the fore here. It enables locally emission-free, $CO_2$-free and low-noise operation, combined with the lowest energy consumption of all compared powertrains.

Despite the high energy efficiencies of the BEV and $H_2$-FCEV, they are not suitable for all applications. This is especially true for vehicles with multiple auxiliary power outputs or applications that do not have access to high-purity hydrogen or corresponding charging infrastructure. For these applications, the robust $H_2$-ICE powertrain or, if necessary, a highly efficient Diesel drive is the optimal $CO_2$-free or low-$CO_2$ solution.

The analysis of the WtW-$CO_2$ balances in Fig. 4 shows that the blue hydrogen pathway is $CO_2$ favourable in the heavy-duty vehicle as well, if the $H_2$ generation in Germany with ordinary electricity mix is assumed. Furthermore, a vehicle with $H_2$ combustion engine with blue hydrogen would offer about 45% lower $CO_2$ emissions (WtW) than a diesel powertrain.

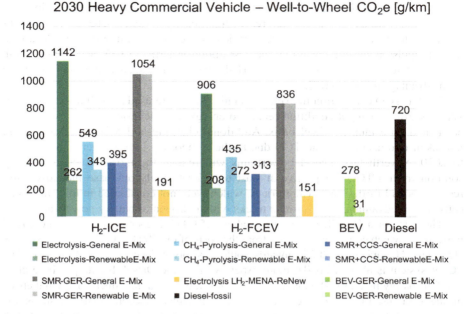

**Fig. 4.** $CO_2$ equivalent emissions WtW for heavy duty commercial vehicles with general and renewable electricity mix in 2030

For 2030, centralized generation of hydrogen with renewable electricity mix is foreseeable, which significantly reduces equivalent emissions for the electrolysis and methane pyrolysis processes. With blue hydrogen, the two hydrogen powertrains produce an additional $CO_2$ emission of 35 to 117 g/km compared to a battery vehicle charged with ordinary electricity mix. Furthermore, it can be seen in Fig. 4 that with renewably produced MENA hydrogen, the equivalent WtW-$CO_2$ emissions can be between 87 and 127 g/km lower than for a battery vehicle charged with ordinary electricity mix, depending on the powertrain system. Only if the BEV is also charged locally with real renewable electricity the WtW-$CO_2$ emissions be reduced to values of about 30 g/km, which is unattainable for any hydrogen application. It can also be stated, that already in TtW the $H_2$-FCEV has an advantage in comparison to the $H_2$-ICE.

The basic influences from the WtW-analysis remain also for the CtG view. Thus, the use of grey hydrogen without CCS is still to be avoided from a $CO_2$ point of view. The combination of those fuels with any hydrogen powertrain is neither with WtW nor CtG balancing better than a fossil-fuelled diesel. Assuming a target range of 800 km, the BEV systems have a required battery capacity of more than 1100 kWh, which has a significant impact on the $CO_2$ footprint. However, since the heavy-duty vehicles have much higher mileage compared to the LCV and passenger cars, the $CO_2$ impact from the large battery systems is relativized over the lifetime. Depending on the electricity mix used to charge the BEV, the hydrogen powertrains may have higher or lower $CO_2$ emissions. If the battery vehicles are charged with ordinary electricity mix, the CtG-emissions increase to 373 g $CO_2$e/km. Only with domestically produced blue hydrogen and FC-propulsion this $CO_2$ level can be undercut for the same electricity mix. In contrast, MENA hydrogen can reach significantly reduced CtG-$CO_2$ emissions compared to BEV's. Only if the battery vehicles are charged with renewable electricity, these propulsion systems would be the best option in terms of $CO_2$ according to both WtW and CtG balancing. Looking at the $H_2$-FCEV it can be seen, that also in CtG there is an advantage to the $H_2$-ICE.

For the TCO calculation the vehicle retail prices were used as well as for the other vehicles. Another input condition relates to energy and fuel consumption, which was calculated on a tank-to-wheel basis. Additional charging and refuelling losses were not taken into account in the TCO due to the technology improvements assumed up to 2030. Nevertheless, Diesel powertrains will remain the most attractive mobility solution from a TCO perspective at least until 2030. Only purchase incentives, toll reductions and further increasing prices, which were not considered in this study, can cause a cost advantage for long-haul trucks with hydrogen or battery powertrain.

The price of a $H_2$-ICE powertrain is about 45,000 € higher compared to the Diesel powertrain, which comes in particular from the costs of the 700 bar $CGH_2$ storage system. Here, an $LH_2$ tank from 2025 onwards could possibly reduce the cost difference. The FCEV system is about 78,000 € more expensive than the Diesel, but its purchase price will fall sharply by 2030. The main reason for this are the scaling effects that will then set in due to the use of passenger car fuel cells in a modular concept. In BEV systems, the powertrain costs are mainly driven by the battery size. Due to the range-scaled 1110 kWh battery, the powertrain costs are about double than the Fuel Cell System.

The TCO curves shows an advantage of 3–5 €ct per km for the $H_2$-ICE until 2025. This advantage for the $H_2$-ICE is expected to reverse into a TCO advantage for the $H_2$-FCEV, depending on the rate of decrease of the prices for the FC-systems. The timing of this reversal will also be partly determined by the price of hydrogen over the period 2025–2035. In total, $H_2$-FCEV and BEV are on a similar TCO level in the period 2025–2030. Cheap electricity (<20 €ct/kWh) and a hydrogen price >6 €/kg could give the BEV a benefit in TCO.

In the long-haul sector, the $H_2$-ICE powertrain must go into series production as early as possible in order to be able to leverage a high $CO_2$ reduction potential in the short term at acceptable additional costs compared to Diesel systems. The presumably rapidly decreasing price of the FC powertrains, as well as the further consumption potential, will lead to a widening of the TCO gap between $H_2$-ICE and $H_2$-FC powertrain after 2025. The main focus here is on tank technology, the $LH_2$ technology will offer a significant cost potential. An $H_2$-ICE with mild hybrid and waste heat recovery could close the efficiency gap to the $H_2$-FC powertrain to some extent. The $CO_2$ analysis for WtW and full life cycle shows that hydrogen powertrains can undercut fossil powertrains by a factor of three to four. The hydrogen powertrains are also competitive with the BEV. In terms of $CO_2$, and depending on the electricity mix, both the $H_2$-ICE and the $H_2$-FC even offer a significant potential compared to the BEV powertrains. The TCO balance of this constellation in the period 2025–2035 will depend on how fast the prices for the FC-systems will fall and how expensive hydrogen can be produced in large scale. For heavy-duty regional delivery tasks, the BEV systems will become more prevalent. Special applications also will benefit in the long term from the largely infrastructural independence of a highly efficient Diesel powertrain and the robustness of ICE's in general ($H_2$ and Diesel).

## 6   H$_2$-Infrastructure and Life Cycle Footprint

The results of the Life Cycle Assessment (LCA) are shown in the Figure 5, where the left column shows the results for the electricity mix for 2030 (220 g CO$_2$e/kWh) and the right column the figures for pure renewable electricity. The allocation of energy consumption for compression and liquefaction is not completely clear-cut: In this diagram, only compression for road transport is listed separately; any compression work at the filling station or for grid transport is included in the respective categories.

The following core statements can be derived

1. The production of hydrogen by electrolysis from (partly) fossil-derived electricity is – as expected – ecologically nonsensical.
2. The CO$_2$ emissions of hydrogen produced by SMR are – as expected – independent of the electricity mix.
3. Turquoise hydrogen (methane pyrolysis) is comparable to blue hydrogen (SMR + CCS) not only in terms of cost, but also in terms of CO$_2$ emissions. Production with non-renewable electricity is quite conceivable – unlike electrolysis.
4. Transport by ship (excl. liquefaction) is significantly more energy efficient than by long-distance pipeline (not shown separately). This always leads to an advantage for LH$_2$ import. If the energy consumption for regular recompression in the pipeline is covered by fossil electricity, this is reflected in significant additional emissions.

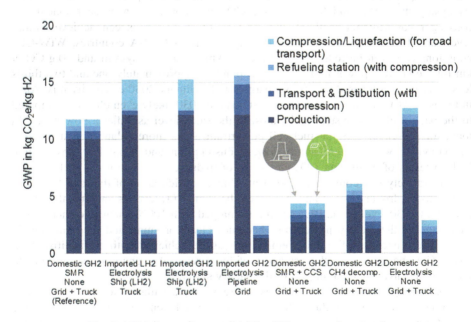

**Fig. 5.** Global warming potential for different supply scenarios

44   C. Danzer et al.

Analyzing Fig. 5 more in detail, several viable paths exist to achieve acceptable $CO_2$ emissions: In the case of blue (SMR + CCS) and green (import) hydrogen, environmental and economic attractiveness coincide, with $LH_2$ import preferred over $GH_2$ import. Green hydrogen produced in Germany by electrolysis is comparatively expensive but has among the lowest $CO_2$ emissions and is a possibility in certain special cases. For methane pyrolysis, there is in no case a unique selling point under the assumptions made; turquoise hydrogen is therefore always comparable with blue. However, this strongly depends on the costs for $CO_2$ disposal and the prices that can be achieved with carbon; a final statement is difficult. However, it can be assumed that methane pyrolysis (as a plasma reactor with renewable electricity) will increasingly replace production by steam reforming, but only on a large scale after 2030.

Furthermore, should $CO_2$ disposal become a significant cost factor, it can be assumed that the production of blue hydrogen by means of SMR + CCS will take place in the immediate vicinity of $CO_2$ storage sites in the future (i.e. abroad); such a scenario was not considered here.

# 7 Conclusion

Hydrogen powertrains offer a huge potential to contribute economically and ecologically to a sustainable mobility sector. In 2030 blue hydrogen produced with 220 g $CO_2$e/kWh electricity emissions offer a WtW-$CO_2$ potential of 54% ($H_2$-ICE) and 72% ($H_2$-FC) compared to diesel powertrains in the passenger car segment. Analysing the WtW- and full life cycle $CO_2$ emissions, it can be seen that in the passenger car segment, both $H_2$-ICE and $H_2$-FC powertrains can achieve similar levels as BEV's. With green hydrogen imported from MENA countries, WtW-$CO_2$ emissions can even be reduced to 18 g $CO_2$e/km for the FC system and 30 g $CO_2$e/km for the $H_2$-ICE powertrain. This corresponds to approximately one and two thirds less emissions, compared to a BEV charged with ordinary 2030 electricity mix. Also, in terms of TCO, both $H_2$ propulsion systems are 2030 likely even cheaper compared to the battery electric system and almost at the same level as a diesel powertrain. In total, for passenger cars the fuel cell powertrain is the more balanced system, as it can convince with a significantly lower consumption and a smaller tank system both in terms of equivalent $CO_2$ emissions of hydrogen production, as well as with a comparatively low component footprint. Under consideration of the total life cycle emissions, hydrogen powertrains can reach the same $CO_2$ level as an equivalent BEV already in 2030, even if the BEV would be charged with 100% renewable energy. The advantage of the battery powertrains becomes clear for small and medium vehicle sizes, where excellent life cycle emissions can be achieved with competitive total costs. The challenges are the provision of local renewable charging energy and to improve the environmental impact of the battery system.

For the light commercial vehicles, there is a particular potential for the $H_2$-ICE powertrain, which reduces the efficiency-related consumption gap to the FC powertrain by means of a high specific utilization. Moreover, the power scaling between passenger cars and light commercial vehicles can be achieved costlier with a combustion engine than this is possible with the more complex FC powertrain system. In 2030, the $H_2$-ICE powertrain operated with blue hydrogen produced with ordinary electricity mix causes nearly the same WtW-$CO_2$ emissions compared to a BEV with the same range and the same electricity mix. A further $CO_2$ and cost potential is possible with imported MENA hydrogen, where in terms of TCO the $H_2$-ICE is even the best hydrogen powertrain. Altogether, both the FC and the $H_2$-ICE powertrains are interesting alternatives to fossil and battery electric powertrains under all evaluation criteria. Based on the premises made, there is a slight advantage of the $H_2$-ICE compared to the FC systems. Especially the high costs and the $CO_2$ footprint of the BEV systems show that for LCV's a range of about 250 km should not be exceeded in order to achieve a balance between costs, payload and $CO_2$ footprint.

The heavy-duty truck sector is characterized by a high diversity of use cases and must contribute a significant share to the $CO_2$ reduction of the transportation sector. In the long-haul truck sector, vehicles with $H_2$-ICE can become the pioneers of a hydrogen infrastructure. Already in 2024, this technology will be ready for series production and thus contribute to a significant $CO_2$ reduction in the fleet mix. After solving technical challenges, especially regarding robustness, $H_2$-FC trucks can be seen as the preferred hydrogen technology in the long term. A highly efficient Diesel is attractive to leverage $CO_2$ potentials in combination with zero emission vehicles in the fleet in the short term and offers potential in the long term only if legislation considers e-fuels as zero emission fuels. Battery electric trucks will become widely accepted especially in the regional delivery sector, for long-haul applications the technological disadvantages and limitations of use (range, payload, charging infrastructure) will lead to a breakthrough of hydrogen technologies. Special applications also will benefit in the long term from the largely infrastructural independence of a highly efficient Diesel powertrain and the robustness of ICE's in general ($H_2$ and diesel). The heavy-duty vehicle sector can play a key role in the establishment of the hydrogen infrastructure: due to the plannable routes and refueling processes in central hubs, fleet operators have the opportunity to build their own hydrogen infrastructure. This speeds up the infrastructure extension and allows the regenerative hydrogen production at favorable prices.

# References

1. Wukisiewitsch, W., Danzer, C., Semper, T.: Systematical development of sustainable powertrains for 2030 and beyond. MTZ Worldw. **81**(2), 30–37 (2020)
2. Danzer, C., Kratzsch, M., Vallon, M., Günther, T.: Fleet Powertrain 2025 – $CO_2$- and cost-optimized modular powertrains. MTZ Worldwide. **79**(2), 36–41 (2018)
3. Umweltbundesamt: Wege in eine ressourcenschonende Treibhausgasneutralität – RESCUE. Dessau-Roßlau (2019)

# Requirement and Potential Analysis of Load Profile Prediction Algorithms

Lukas Schäfers[1]($\boxtimes$), Pascal Knappe[1], Rene Savelsberg[2], Matthias Thewes[2], Simon Gottorf[2], and Stefan Pischinger[1]

[1] Chair of Thermodynamics of Mobile Energy Conversion Systems, RWTH Aachen University, Aachen, Germany
{schaefers_l,kanppe,stefan.pischinger}@tme-aachen.de
[2] FEV Europe GmbH, Aachen, Germany
{savelsberg,thewes,gottorf}@fev.com

**Abstract.** In the following paper, requirements for long-term power demand prediction algorithms are formulated, that are needed for a successful implementation of prediction-based driving functions and for enabling their potential benefits in series production vehicles. To this end, influencing factors that affect the load profile when driving on a given route are identified and examined for the impact of their availability and information quality. Load predictions of varying accuracy – i.e., in presence or absence of certain information or even misinformation – are generated and analyzed for their potential benefits when applying a range estimation algorithm for battery electric vehicles and a predictive control strategy for hybrid electric vehicles using discrete dynamic programing (DDP). It is demonstrated that the prediction quality has a significant impact on the benefit of these strategies. When the prediction accuracy is low, the energy demand using a DDP strategy that promises globally optimal control may even be increased compared to rule-based strategies. It is also shown that different predictive applications have different requirements on their prediction quality. The results thus provide an important contribution to the improvement of load prediction algorithms and to the introduction of long-term predictive functions to production vehicles in future.

**Keywords:** Predictive Driving · Load Profile Prediction · Electronic Horizon · Residual Range Estimation · Hybrid Management

## 1 Introduction

Various applications in vehicles can be optimized predictively in case the future load and energy demand required to drive a particular route are known in advance. Globally optimized energy management for hybrid powertrains, thermal management and residual range estimation in battery electric vehicles or the heating strategy of the exhaust aftertreatment system and predictive onboard diagnosis (OBD) planning in

© Der/die Autor(en), exklusiv lizenziert an Springer Fachmedien Wiesbaden GmbH, ein Teil von Springer Nature 2023
A. Heintzel (Hrsg.): ATZLive 2022, Proceedings, S. 46–61, 2023.
https://doi.org/10.1007/978-3-658-41435-1_5

conventional vehicles are just a selection of long-term prediction-based functions. The potential of these functions to further optimize powertrain efficiency or enhance the driving experience, e.g. by reducing range anxiety, has been demonstrated in various publications. Most of these publications, however, focus on the optimisation of the respective control algorithms, and results are demonstrated using previously known or common standard driving cycles [1]. The prediction of the load profile, especially over longer horizons, is in contrast often neglected or subject to assumptions and simplifications. In series production battery electric vehicles for example the predicted residual range often still depends simply on the average past energy demand instead of considering the route ahead (e.g. [2]).

To enable the benefits of predictive functions in production vehicles also accurate and robust forecasting algorithms are needed. The challenges here are usually their dependence on real-world data that is prone to uncertainties and available to limited extent at the beginning of the trip. To be still able to develop precise and robust prediction algorithms an analysis of the impact of these uncertainties is mandatory. Also the target accuracy of the prediction must be known so that downstream predictive functions still provide benefit. The following sections address these questions, analyze what factors affect the accuracy of the prediction and how these can be incorporated into the prediction approach. To do so the remainder of this paper is organized as follows:

First, related research papers and publications are analyzed that focus on predictive algorithms and conclude on requirements on their prediction accuracy. In section three the analysis approach followed in this paper is introduced, starting with the analysis of an energy and load prediction algorithm, the uncertainties in the prediction as well as their impact on the prediction results. Then, the two exemplary long-term predictive functions are described based on which the requirements on prediction accuracy are evaluated. In section four the simulation results are presented and discussed. Conclusion and outlook are given in section five.

## 2 Related Publications

As stated in the comprehensive survey on prediction algorithms for predictive hybrid energy management functions of Zhou, Pera et al., in the past there has been little research published on prediction algorithms compared to the research conducted on the predictive energy management functions themselves [1]. Also recent publications in this field assume a given load profile or known driving cycle. In [3] for example the authors present a novel hybrid control strategy based on iterative dynamic programing trying to solve the "curse of dimensionality" problem of conventional DP-based strategies. The results are presented assuming the previously-known FTP-72 driving cycle. Also in [4], in which the authors expand a model-predictive control (MPC) based strategy for plug-in hybrid electric vehicles by thermal constraints to further optimize energy consumption, the applied FTP-72 cycle is assumed to be known. The DP-approach with receding horizons in [5] is presented using the WLTC.

48  L. Schäfers et al.

Authors in [6] in contrast try to solve the problem of prediction uncertainties by adopting the optimisation accordingly. They introduce a stochastic dynamic programming approach that decreases energy demand of unknown driving cycles depending on a single tuning parameter while losing global optimality. In [7] a dynamic programming algorithm for optimizing hybrid vehicle control is combined with a load profile prediction using neural networks. The prediction accuracy of the algorithm or its impact on efficiency benefit of the predictive control strategy, however, is not subject of the publication.

Besides publications on predictive hybrid controls there are also publications focusing on the load profile prediction itself. [1] gives an overview of different prediction approaches, their prediction accuracy and robustness, influencing factors, possible applications and even factors causing miss-prediction. What is not yet covered by the introduced publications is the impact of the different prediction accuracies on the results of downstream predictive functions. Therefore, these questions are subject of the following expositions.

## 3  Concept & Analysis Approach

To allow the analysis of accuracy requirements for load profile forecasting algorithms, first the system to be predicted is introduced and a simplified signal flow of a model-based prediction approach is presented. Then, uncertainties and sources of prediction errors are identified by analyzing the different input data of the algorithm. Once the effects of uncertainties on the prediction results are known, the impact on the benefit of downstream predictive functions can be analyzed in a simulation study.

### 3.1  Load Profile Prediction

Most long-term predictive functions rely on the load profile or corresponding energy demand required to drive a given route, either directly to e.g. optimize the torque split of a hybrid electric vehicle or indirectly for estimating the thermal load of components and optimizing the thermal management accordingly. To ensure that the vehicle can be operated using predictive functions during most of the trip, the prediction information must be available shortly after departure or even prior to the trip. A simplified overview and signal flow of a model-based prediction algorithm for such load profile is depicted in Fig. 1.

The total vehicle load is the sum of the power required for traction $\hat{P}_{\text{Trac}}$ and power required for auxiliaries $\hat{P}_{\text{Aux}}$ like the Heating, Ventilation and Air Conditioning (HVAC) system, multimedia and other board electronics. The required power for traction depends on one hand on the vehicle resistances $\hat{F}_{\text{R}}$, i.e., the power needed for driving the vehicle in its current state as requested by the driver, and on the other hand on the losses that are generated when providing this power via the powertrain components. Depending on the topology of the drivetrain, losses are generated as friction losses, electrical losses and thermal losses in the different components. Given a certain vehicle speed, acceleration and corresponding driving resistance forces $F_R$, the prediction of these losses is mostly depending on component-related coefficients and their thermal states, and are less depending on unpredictable environmental factors. They can be modelled using simple drivetrain models based on efficiency coefficients or maps and uncertainties are relatively small. The relevant driving resistances $F_R$ and related wheel torque result from the equilibrium of forces in longitudinal direction of a vehicle and can be divided into acceleration resistances (rotational and translational) $F_{a,tot}$, slope resistance $F_{Slop}$, rolling resistance $F_{Roll}$ and air resistance $F_{Air}$ (see Eq. 1). The cornering resistance due to lateral forces acting in longitudinal direction is neglected in this case.

$$F_R = F_{a,tot} + F_{Slop} + F_{Roll} + F_{Air} \qquad (1)$$

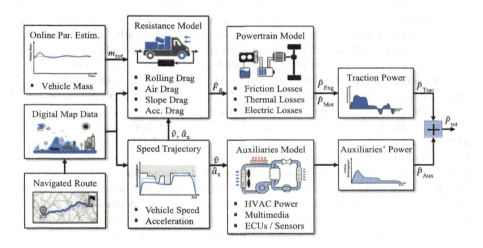

**Fig. 1.** Simplified signal flow of a model-based prediction of the load profile of a vehicle driving on a certain route

A common way to estimate driving resistances is the following semi-empirical vehicle resistance formula (Eq. 2) with the total vehicle weight $m_{tot}$, reduced rotational inertia $\theta_{red}$, dynamic tire radius $r_{dyn}$, vehicle acceleration a, gravitational acceleration g, road slope $\alpha_r$, rolling friction coefficient $f_R$, air density $\rho$, air drag coefficient $c_w$, vehicle cross-span area A and vehicle speed v [8, 9].

$$F_R = \left( m_{tot} + \frac{\theta_{red}}{r_{dyn}^2} \right) \cdot a + m_{tot} \cdot g \cdot \sin\alpha_r + m_{tot} \cdot g \cdot \cos\alpha_s \cdot f_R + \frac{\rho}{2} \cdot c_w \cdot A \cdot v^2. \qquad (2)$$

Eq. 2 shows dependencies on less well-predictable influencing factors. Whereas tire radius and rotational inertia can be considered as known parameters given a vehicle speed and appropriate gear selection, the vehicle mass may vary from trip to trip with the number of passengers and cargo load. Therefore, it is a source of unrtainty

50 L. Schäfers et al.

and should be predicted at the beginning of each trip. In case the route is known, road slope and road conditions can be obtained from digital map data and weather information like wind speed and air density are available from weather services. The friction and air drag coefficient, usually determined based on coastdown tests, may instead vary depending on these weather and road conditions [10]. Most importantly, however, is the influence of the vehicle speed and acceleration. Compared to the other variables in Eq. 2, they have the largest range of values and the quadratic dependence on vehicle speed results in the air drag accounting for the largest fraction of the total resistances at higher vehicle speeds [9].

Although controlled by the driver, vehicle speed and acceleration depend on diverse impact factors that are difficult to predict. The acceleration and deceleration behaviour, i.e., the aggressiveness or passivity of driving style, is relatively freely chosen by the driver within the boundaries of powertrain performance. The vehicle speed instead does not solely depend on the driver's preference. The driver must consider traffic rules like speed limits, traffic conditions, traffic light signals and stop signs, road conditions, curvatures, visibility conditions and other aspects. Some of these factors are static and predictable, such as speed limits and curvature radius, some are highly dynamic and nondeterministic like traffic speed or the status of traffic lights. Therefore, there are many sources of uncertainties in the prediction of speed trajectories and, hence, of the vehicle resistances.

The auxiliaries' consumption $\hat{P}_{Aux}$ also depends on both, static and dynamic aspects. The system design, i.e. the number of electrical components like vehicle sensors and multimedia system components, but also vehicle speed, the passenger's preferences and weather conditions like environmental temperature and solar radiation determine the auxiliaries' consumption. Due to the complexity and variety of different auxiliary systems in this case the corresponding load is not modelled but represented by a measured load. It should be noted, that depending on the type of vehicle the auxiliaries constitute a significant fraction of the total power. While in conventional vehicles heating can be provided using waste heat of the engine and mostly cooling affect the energy demand, in battery electric vehicles the HVAC system can account for up to approximately 40% of the total power (e.g. [11, 12]).

### 3.2 Prediction Variations

Summarizing the preceding discussion, vehicle mass, auxiliaries' consumption and mainly vehicle speed are prone to uncertainties and influence the load profile required for driving on a given route. Therefore, these aspects are in focus of the following analysis of their prediction requirements. To demonstrate the impact of predictions of varying accuracy, two downstream predictive functions are applied: a residual range prediction algorithm for battery electric vehicles and a globally-optimised control strategy based on backward calculating Discrete Dynamic Programming (DDP) for hybrid electric vehicles. The baseline for comparison is defined by the maximum benefit of the two predictive strategies, which is obtained by assuming full knowledge of the load profile and comparing the results to the alternative rule-based strategy, respectively. This best-case scenario cannot be further improved by the forecasting algorithm but only by the predictive optimisation strategies themselves. The influence of the prediction algorithms therefore becomes quantifiable by varying the prediction accuracy while monitoring the impact on the initial benefit. A reduction in prediction accuracy will lead to a reduction of the residual benefit and requirements on the prediction accuracy can be derived.

The full knowledge scenario is based on a real-world measurement taken by a light duty vehicle of $m_{tot} = 3000$ kg weight driving in urban and rural area. The measured vehicle speed is shown as solid grey line in Fig. 2, while the measured auxiliaries power demand and road slope are depicted in the bottom diagram. The stepwise variation of the prediction of this measurement contains the following stages. First, the predicted speed levels along the route are varied starting with a rough estimate, which would result from considering just the speed limits of the traveled route (blue solid line in Fig. 2). More precise estimations would result from considering traffic speed information, driver behaviour or weather conditions, represented by medium accurate and detailed constraints (solid black and dashed red line in Fig. 2). For all predictions a motion equation based differentiable speed profile is fit into the constraints, as exemplified for the medium speed constraints (dashed black line).

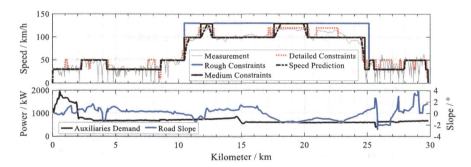

**Fig. 2.** Measured vehicle speed, varying levels of speed constraints and derived predicted speed profile (top) and auxiliaries demand and road slope of the example measurement (bottom)

Next to the speed levels also the predicted stops during the trip are varied. Using the medium speed constraints as base, two more variants are added: one with assuming the actual stops as they have been measured to be known (black line in the left diagram of Fig. 3) and one with also all other potential stops along the route resulting from traffic lights, stop signs or turns, even though during the measurement the vehicle did not stop at this point (red dotted line in left diagram of Fig. 3).

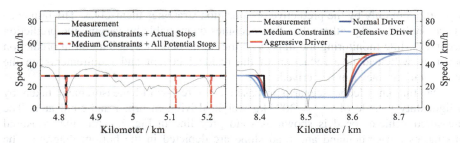

**Fig. 3.** Actual and potential stops along the route (left subplot) and different driving styles leading to different speed profile prediction results (right subplot)

Finally, the acceleration and deceleration behaviour is modified, resulting in a normal, aggressive and passive driving style (right diagram of Fig. 3). With differed vehicle mass (too low and too much weight compared to actual weight), and the auxiliary's power being neglected in total ten variations of prediction accuracy are analyzed as listed in Table 1. It shall be noted that this list doesn't claim to be complete but shall rather demonstrate the entire approach of analyzing the requirements for forecasting algorithms.

**Table 1.** Scenarios and prediction accuracies

| #  | Speed Constraints | Driving Behaviour | Considered Stops   | Vehicle Mass | Auxiliaries' Power |
|----|-------------------|-------------------|--------------------|--------------|--------------------|
| 1  | Rough             | Normal            | Neglected          | 3000 kg      | As measured        |
| 2  | Medium            | Normal            | Neglected          | 3000 kg      | As measured        |
| 3  | Detailed          | Normal            | Neglected          | 3000 kg      | As measured        |
| 4  | Medium            | Normal            | As measured        | 3000 kg      | As measured        |
| 5  | Medium            | Normal            | All potential stops| 3000 kg      | As measured        |
| 6  | Medium            | Aggressive        | As measured        | 3000 kg      | As measured        |
| 7  | Medium            | Passive           | As measured        | 3000 kg      | As measured        |
| 8  | Medium            | Normal            | As measured        | 2500 kg      | As measured        |
| 9  | Medium            | Normal            | As measured        | 3500 kg      | As measured        |
| 10 | Medium            | Normal            | As measured        | 3000 kg      | Neglected          |

### 3.3 Use Case Predictive Hybrid Control Strategy using Discrete Dynamic Programming

For the impact analysis of the varied prediction results on a predictive hybrid control strategy a Volkswagen Crafter prototype, developed by FEV Europe GmbH in cooperation with Volkswagen AG, is used. It is the same vehicle that also the measurement was taken with. As shown in Fig. 4, the conventional drivetrain has been converted into a plug-in parallel hybrid vehicle in P0-P2 configuration. The vehicle is operated as a P2 hybrid while the belt starter generator (BSG) is used solely for engine starts and charging while stationary, if necessary [13].

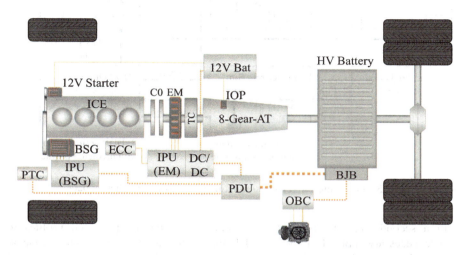

**Fig. 4.** Overview of the investigated P0P2 PHEV vehicle model

As vehicle model a forward-oriented Matlab/Simulink model with a simplified, mostly map-based, representation of all controllers and physical components as well as a driver model is used. The implemented baseline operating strategy is a rule-based strategy that is calibrated for low fuel consumption. This strategy is purely causal and does not take into account the future load profile. The predictive strategy, an optimisation-based backward calculating DDP approach, in contrast does consider the future load profile, globally optimizing the torque split between internal combustion engine and electric motor and controlling engine start and stop events. A simplified simulation procedure for testing the DDP strategy with predicted load profiles of varied accuracy is shown in Fig. 5. First of all, a simulation of the rule-based operating strategy using the measured speed and load profile is performed. The final battery state of charge (SOC) serves as target SOC for the DDP and the efficiency results as baseline comparison value.

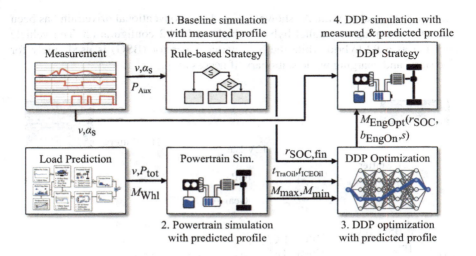

**Fig. 5.** Procedure for vehicle simulation with falsified prediction data

In a second step, the powertrain is simulated using the predicted profiles to estimate necessary input data for the DDP like temperatures of the engine coolant, engine and transmission oil, chosen gears and torque limits of the electric motor and the combustion engine. Afterwards the DDP optimisation is executed, calculating backwards from the final SOC of the baseline simulation. The result of the optimisation is a three-dimensional matrix, which contains the optimum combustion engine torque depending on the current SOC, the current travelled distance as well as the current engine operating state – i.e. combustion engine on or off. Unlike normally, the result of the DDP optimisation in this case is a function of distance rather than time because the travel times and the times the vehicle is at a given point are different, when the predicted speed profiles are varied [14]. Finally, the simulation of the DDP strategy can be performed using the measured speed and load profile as driving cycle with the optimum engine torque matrix based on the predicted profile as operating strategy. This allows to simulate the case that at the beginning of a trip a non-exact prediction is made based on which the operating strategy is optimized (leading to non-optimal results). This may lead to a final SOC that does not perfectly match the target SOC. In this case the fuel consumption is corrected by converting the SOC difference into a fuel mass using the fuel equivalent for the investigated result. By applying different predictions the influence of the prediction accuracy can hence be quantified and requirements can be derived.

### 3.4 Use Case Residual Range Prediction

Provided a predicted load profile, residual range prediction algorithms for battery electric vehicles are less complex compared to the DDP strategy. As shown in Fig. 6, there is no optimisation subsequent to the prediction. Instead, the predicted load profile is simply integrated to estimate the required energy demand and then subtracted from the energy stored in the high-voltage battery indicated by the effective current SOC.

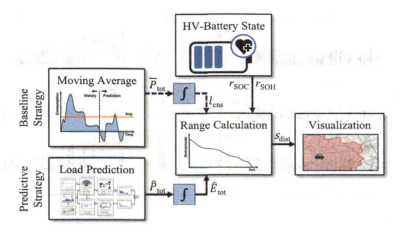

**Fig. 6.** Simplified signal flow of a predictive residual range prediction algorithm in comparison to a non-predictive moving average approach

The challenge of developing precise residual range prediction algorithms lays – apart from accurate battery state estimation which is not subject of this paper – in the prediction of the energy demand itself. The requirements on the prediction algorithm therefore may differ from those of e.g. the predictive control strategy for hybrid electric vehilces. The temporal accuracy of the prediction of load peaks for example is of secondary importance due to the low-pass nature of the integrator, whereas the average load must be predicted accurately to avoid accumulated errors.

The benefit of a residual range prediction algorithm cannot be measured in efficiency improvement. Therefore, the prediction accuracy itself is taken as indicator for functional benefit. The benchmarking strategy in this case is a simple moving average of the past energy demand as it is implemented in many state of the art production vehicles (e.g. [2]). In this strategy, the energy demand of the past 20 kilometers is averaged and extrapolated into the future assuming constant energy demand for the route ahead. Since the residual range at the end of the trip depends also on future trips it is an unsuitable value for the comparison. Instead, the predicted final SOC at the end of the measured trip is evaluated for both strategies in comparison with the measured SOC.

## 4 Results Discussion

### 4.1 Predictive Hybrid Control Strategy

The following figure (Fig. 7) shows exemplary results for the rule-based strategy, the DDP simulation with full knowledge and the DDP simulation with a rough estimated profile, which corresponds to scenario #1 in Table 1. The upper three diagrams indicate the chosen operation modes of the corresponding strategies while the lower diagram depicts the SOC trajectories. For the rule-based strategy the SOC is constantly close to the target of $r_{SOC,tar} = 0.25$.

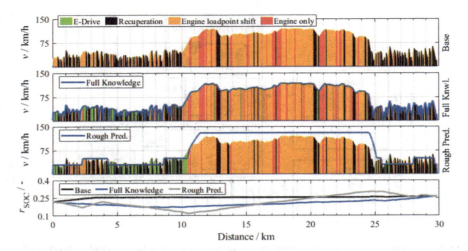

**Fig. 7.** SOC trajectories and powertrain operation modes for the rule-based strategy, the DDP strategy assuming full knowledge and assuming a rough prediction

The DDP with full knowledge instead uses much more electric driving, mostly when there is a low load demand, and charges the battery via load-point-shifting during the highway phase between kilometre 10 and 25. Overall this results in a fuel consumption benefit of 8.4% compared to the baseline strategy, as shown in Fig. 8, where the absolute and relative fuel consumptions of all scenarios are compared. The result of prediction #1 is more comparable to the DDP with full knowledge in terms of the phases of electric and hybrid driving. The fuel consumption, however, is about 6.14% higher compared to the rule-based strategy. A closer look at the operation modes and the driving profile with rough prediction reveals that hybrid driving is mostly active when the predicted speed and hence the load is relatively high. The SOC trajectory, however, shows that despite this similarity of the operating modes the charging strategy is quite different. With the optimisation result of prediction #1, charging is clearly more aggressive, and in some cases also taking place at relatively low speeds, where the benefit of load-point-shifting is rather small. This means that the combustion engine's efficiency is not optimal and, above all, the electrical losses are comparatively high due to high electrical currents. As a result, the rough prediction is not sufficient to reach the goal of a lower fuel consumption as the DDP with full knowledge did.

With increased prediction accuracy, as in prediction #2 and #3, the fuel consumption drawback can be reduced (see Fig. 8). Assuming the medium-detailed speed constraints as base for prediction (#2) the fuel consumption is 1.18% lower than when using rule-based strategy, but still 7.22% above the result with full knowledge. With the detailed prediction the fuel consumption is still 5.91% above best case scenario and 2.48% below the baseline. This demonstrates the major influence and importance of predicting an accurate speed trajectory when applying prediction-based hybrid control strategies using DDP.

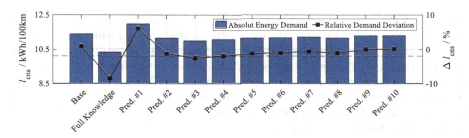

**Fig. 8.** Relative and absolute fuel consumption of the DDP strategy using different prediction accuracies

Considering vehicle stops further improves the fuel consumption (#4). The improvement, however, is only about 0.72%. In terms of time, acceleration and deceleration phases (i.e. stops) only have a small share in the analysed measurement with respect to the overall profile. For drive cycles with long phases of highway driving therefore incorporating the information of potential stops has less effect. For the same reason the misinformation of potential stops (#5) and the variation of the driver behaviour in scenarios #6 and #7, which is only represented as a difference in the vehicles' acceleration and deceleration, in this case have no big impact on the results. Also differentiating the vehicles mass, as in the predictions #8 and #9, has a rather low influence. The reason for this, however, is different from that for the small impact of the driving behaviour. Despite its effect on acceleration and deceleration phases, vehicle mass also affects the slope drag. Since the elevation profile of the considered route is relatively flat and there is almost no increase in altitude over distance also the influence of vehicle mass is low. With more acceleration phases and non-zero altitude gain, the impact of an inaccurate vehicle mass, however, is expected to increase. Last but not least in scenario #10 the impact of missing information regarding the auxiliary power shows a significant impact. This is caused by the fact that the axillary units are consuming power over the whole time which has an influence on the optimization results along the entire route.

## 4.2 Residual Range Prediction

For the residual range prediction use case, an example result of the comparison between the predictive strategy and averaging baseline strategy for the rough speed profile prediction (No. 1 in Table 1) is shown in Fig. 9. The upper left diagram shows again the predicted and measured vehicle speed. The average energy demand per covered distance is compared for both strategies to the measured energy demand in the lower diagram. During the first two kilometers a low accuracy is observed for the baseline strategy, overestimating the energy demand. This overestimation results from the measured peak demand at the beginning of the trip – when distance is not yet covered but power is demanded from the auxiliaries. After 10 km, when the vehicle enters the highway, the measured energy demand increases and the baseline strategy shows underestimation. The predictive strategy in contrast predicts the increased energy demand while overestimating it due to an overestimation of the vehicle speed. Because of the same reason also the final SOC of the predictive strategy is underestimated in the first phase of the trip, as shown in the center diagram. The blue line here represents the final SOC as predicted at each step of the trip considering the actual measured SOC at this step as starting point. Due to decreased length of the prediction horizon and the corresponding decreased uncertainty over time, the predicted final SOC converges towards actual measured final SOC with covered distance. The baseline strategy in contrast overestimates the final SOC at the beginning of the trip, since the later highway drive is not yet taken into account.

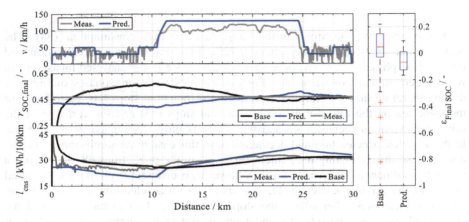

**Fig. 9.** Example result of a predictive range prediction vs. baseline averaging strategy

The boxplot diagram in Fig. 9 shows the deviation between predicted and measured final SOC for both strategies taking one value every $s_{Update} = 100$ m. The absolute median of the predictive strategy is greater than the one of the baseline strategy. The variance and range of values of the deviations, however, are lower for the predictive strategy. It shall be noted, that the magnitude of deviation, and hence its relevance, also depends on the battery's maximum capacity which is in this case $c_{HVmax} = 20$ kWh.

The results of all analyzed prediction steps are shown in Fig. 10. The lower diagram again depicts the deviation between predicted and measured final SOC in form of boxplot diagrams. The upper diagram displays the change in absolute median prediction error taking prediction #4 (medium constraints, normal driver, stops as measured) as base line. Increasing the prediction quality of the speed constraints from rough to medium (#1 vs. #2) results in both, improved median value and decreased variance in predicted values. The final SOC is now overestimated instead of underestimated, since the strong overestimation of the speed profile is mitigated in prediction #2. Also more detailed speed constraints (#3) and the integration of stops (#4) further improves the results. Even the misinformation of stops that are predicted but not measured decrease the prediction error because the additional assumed power of mispredicted stops counteracts the overestimation of the final SOC.

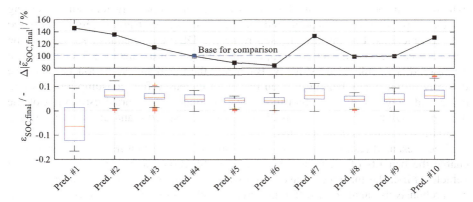

**Fig. 10.** Prediction results for estimating the final SOC using predictions of different accuracies

The driving behaviour has a significant impact on the prediction results. Simulating an aggressive driver in prediction #6 decreases the median prediction error and has the same effect as the misinformation. The defensive driver in contrast leads to a decreased prediction accuracy with the average median being approximately 38% higher than the same prediction with average driving style (#4). The vehicle mass (both, underestimation #8 and overestimation #9) in this case has no effect on the prediction results due to the flat elevation profile and vehicle mass having limited effect on acceleration drag for vehicles with strong recuperation capabilities. A much stronger impact of vehicle mass is expected for driving cycles with a non-zero elevation profile. Prediction #10 in contrast again results in overestimation of the final SOC because of the negligence of the auxiliary's power.

In conclusion it can be stated that already a very rough model-based prediction can outperform the state-of-the art averaging demand forecast when the load profile is drastically changing during the trip. Taking into account information like stops and more detailed speed constraints further improves the results. Impact factors like driving style, auxiliaries demand and probably also vehicle mass, however, should also be estimated carefully to achieve precise forecast results. Obviously, for driving cycles with rather constant power demand in contrast the moving average baseline achieves accurate results.

## 5 Conclusion & Outlook

The impact of prediction accuracy on the benefit of prediction-based driving functions often remains underexplored when predictive functions are introduced. To analyze this correlation, in this paper model-based predictions of varying accuracy are investigated on their influence on a globally-optimized hybrid operating strategy and a range prediction algorithm. Although the presented study cannot be considered exhaustive, it becomes clear that uncertainties in impact factors and resulting inaccurate predictions can lead to undesirable results of downstream prediction-based driving functions. For the investigated route the DDP-based control strategy in the best case can lead to a fuel consumption benefit of approximately 9% compared to a non-predictive rule-based strategy. But when the upstream prediction is imprecise it hardly provides any benefit or even results in increased fuel consumption. Also, in context of range prediction algorithms the importance of accurate estimation of the driving behavior – mainly in terms of chosen vehicle speed – and the often-neglected auxiliaries' demand is demonstrated.

The presented results imply various research directions that need to be pursued to enable the benefits of prediction-based driving functions in future series production vehicles. First, the introduced investigations in this paper need to be continued more comprehensively, analyzing different driving cycles, more strongly varied sampling of impact factors and the statistic distribution of expected benefits or negative consequences. Also other vehicle types and scenarios need to be taken into account. E.g. on routes with higher slopes the vehicle mass is expected to have a much higher impact, especially for heavy duty applications, as their mass can vary in a much wider range. Also more extreme weather conditions, especially cold conditions when auxiliaries can have a much higher impact, can lead to more extreme results.

Apart from these investigations, robust prediction algorithms need to be developed, precisely estimating the vehicle states, monitoring the driver's behaviour and taking into account all relevant environmental information. Also the availability and precision of input data as e.g. live traffic data must be improved. Finally, the downstream prediction-based functions must be designed in a way that the effect of misinformation and incorrect prediction on the function's benefit is limited and the implementation of predictive functions do not lead to decreased performance compared to non-predictive rule-based strategies.

# References

1. Zhou, Y., Ravey, A., Péra, M.-C.: A survey on driving prediction techniques for predictive energy management of plug-in hybrid electric vehicles. J. Power Sources **412**, 480–495 (2019). https://doi.org/10.1016/j.jpowsour.2018.11.085
2. Tesla Inc.: *Model S Owner's Manual.* https://link.springer.com/content/pdf/10.1007%2F978-3-658-36727-5.pdf. Accessed: 22 March 2022
3. Zhu, Q., Prucka, R.: Transient hybrid electric vehicle powertrain control based on iterative dynamic programing. J. Dyn. Sys. Measur. Cont. **144**(2), https://doi.org/10.1115/1.4052230 (2022)
4. Ezemobi, E., Yakhshilikova, G., Ruzimov, S., Castellanos, L.M., Tonoli, A.: Adaptive model predictive control including battery thermal limitations for fuel consumption reduction in P2 hybrid electric vehicles. WEVJ **13**(2), 33 (2022). https://doi.org/10.3390/wevj13020033

5. Polverino, P., Arsie, I., Pianese, C.: Optimal energy management for hybrid electric vehicles based on dynamic programming and receding horizon. Energies **14**(12), 3502 (2021). https://doi.org/10.3390/en14123502
6. Mallon, K.R., Assadian, F.: Robustification through minimax dynamic programing and its implication for hybrid vehicle energy management strategies. J. Dyn. Sys. Measur. Cont. **143**(9) https://doi.org/10.1115/1.4050252 (2021)
7. Arsie, I., Graziosi, M., Pianese, C., Rizzo, G., Sorrentino, M.: Optimisation of supervisory control strategy for parallel hybrid vehicle with provisional load estimate Proc. of AVEC. **4**, 23–27 (2004)
8. Andersen, L.G., Larsen, J.K., Fraser, E.S., Schmidt, B., Dyre, J.C.: Rolling resistance measurement and model development. J. Transp. Eng. **141**(2), 4014075 (2015). https://doi.org/10.1061/(ASCE)TE.1943-5436.0000673
9. Küçükay, F.: Grundlagen der Fahrzeugtechnik. Springer Fachmedien Wiesbaden, Wiesbaden (2022)
10. Descornet, G.: Road-Surface Influence on Tire Rolling Resistance. *ASTM special technical publication*, vol. 1031, *Surface Characteristics of roadways: International research and technologies*. In: Meyer, W.E., Reichert, J. (Eds.) Philadelphia, Pa.: ASTM, 401-401-15. https://www.diva-portal.org/smash/get/diva2:669244/FULLTEXT01.pdf. Accessed: 22. March 2022 (1990)
11. Farrington, R., Rugh, J.: Impact of Vehicle Air-Conditioning on Fuel Economy, Tailpipe Emissions, and Electric Vehicle Range: Preprint. https://www.nrel.gov/docs/fy00osti/28960.pdf
12. Jefferies, D.: Energiebedarf verschiedenerKlimatisierungssysteme für Elektro-Linienbusse
13. Schaub, J., et al.: Electrified efficiency – diesel hybrid powertrain concepts for light commercial vehicles. In: Liebl, J., Beidl, C., Maus, W. (Eds.) Proceedings, Internationaler Motorenkongress 2020, pp. 335–352. Springer Fachmedien Wiesbaden, Wiesbaden (2020)
14. Ambühl, D.: Energy management strategies for hybrid electric vehicles. ETH Zurich (2009)

# Depending on Lithium and Cobalt – The Impact of Current Battery Technology and Future Alternatives

Mareike Schmalz[⊠], Christian Lensch-Franzen, Jürgen Geisler, Amalia Wagner, Thomas Rempel, and Johannes Hüther

APL Automobil-Prüftechnik Landau GmbH, Landau, Germany
`{mareike.schmalz,christian.lensch-franzen}@apl-landau.de`

**Abstract.** As the transition to e-mobility progresses, not only is the number of battery-electric vehicles increasing, but also the individual battery size. In pursuit of ever higher battery capacities and electric vehicle ranges, we experience a rapidly growing demand for certain raw materials. The technology currently used in electric cars is almost without exception the lithium-ion battery which includes various types of battery chemistries and material composites. Required raw materials such as lithium and cobalt have repeatedly been under criticism. To evaluate the new technology, it is important to examine the environmental footprint of the extraction of raw materials as well as to consider ethical and political circumstances. Furthermore, we need to consider which other options exist as new developments in the field of battery cells show alternatives to established materials. A wide range of possibilities opens up, from cobalt-free cells to sodium-based energy storage systems.

As an independent development service provider in the automotive sector, APL deals with current and future technologies on a daily basis and, in addition to ethical arguments, also knows the economic and technical side of how the choice of materials affects the performance of the battery. In this paper, APL gives deeper insights and illuminates the topic from different perspectives.

**Keywords:** Battery · Raw Materials · Cobalt · Lithium · Recycling

## 1 Introduction

### 1.1 Motivation

Electromobility is conquering market shares worldwide, as the number of registrations of battery electric vehicles is steadily increasing. This shift in technology should ensure a smaller environmental footprint and lower greenhouse gas emissions. Further, defossilization of mobility promises a diminished dependence on oil and thus

---

© Der/die Autor(en), exklusiv lizenziert an Springer Fachmedien Wiesbaden GmbH, ein Teil von Springer Nature 2023
A. Heintzel (Hrsg.): ATZLive 2022, Proceedings, S. 62–72, 2023.
https://doi.org/10.1007/978-3-658-41435-1_6

reduced political, ethical and environmental issues. But the new all-electric technology does not come without new raw material dependencies. The necessary elements lithium and cobalt in particular are repeatedly criticized by the press. Both availability and ethical acceptability are questioned. How justified the criticism is, which aspects of the battery raw material chain should be examined more closely and which alternatives need to be found will be discussed below

### 1.2 The Economic Relevance of the Battery Materials

While the e-motor is responsible for the energy-efficient drive and the high dynamics of the electric car, the main criteria in the purchase decision – range, charging time and costs – depend on the battery.

Regarding the cost of the electric powertrain, the battery makes up the lion's share here. Around three quarters of the battery system costs can be attributed directly to the battery cell. These expenses come largely from the material costs, which make up around 60% of the cell costs [1, 22]. This indicates the enormous relevance of battery materials – ethically and economically. With increasing demand for the raw materials, price increases are to be expected.

## 2 Relevant Materials for Lithium-Ion Batteries

### 2.1 Design of Battery Cell

The lithium-ion battery, which currently dominates the traction battery market, contains various materials. It is made up of two electrodes, the anode and cathode. The basic design is shown in Fig. 1.

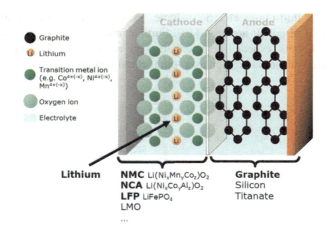

**Fig. 1.** Battery cell design and active materials

The lithium ions, which are responsible for the charge transfer and thus the functionality of the battery, migrate back and forth between the anode and cathode during charging and discharging. The so-called intercalation, takes place in the electrode's active materials, which form the necessary framework for storing lithium. Various materials can be used for this, given that they have the necessary properties for lithium intercalation. Figure 2 shows the dominant elements in battery active materials. Graphite has largely established itself on the anode side, while various material compositions are relevant on the cathode side, for example NMC – nickel manganese cobalt oxide or NCA – nickel cobalt aluminum oxide. Both material composites contain cobalt. Cobalt is used for stability and thus safety of the cell.

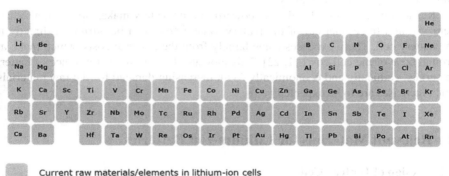

**Fig. 2.** Dominant elements in lithium-ion battery active materials

## 2.2 Raw Material Quantities

Calculated for a mid-range electric vehicle, estimates show that around 6 kg of lithium and 11 kg of cobalt are used [10]. Further raw material shares can be seen in Fig. 3. These amounts, especially for cobalt, vary according to battery size and cell technology used and should be seen as a rough classification.

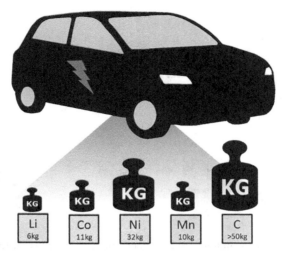

**Fig. 3.** Amounts of raw materials in an average electric vehicle according to [10]

## 3 Lithium

### 3.1 Occurrence and Reserves

At 0.006% of the earth's crust, lithium is slightly rarer than zinc or copper, and slightly more common than tin or lead, although it is more dispersed than lead [2] Nevertheless, as of today, worldwide mining is limited to a few countries [11]. In 2020 the global reserves in the existing mines were estimated by the USGS at around 17 million tons, in 2021 the estimate was increased to 21 million tons [17]. Owing to continuing exploration, identified lithium resources have increased substantially worldwide over the last years. The deposits from continental brines, geothermal brines, from the hectorite mineral, oilfield brines and from the igneous rock pegmatite has been estimated at over 80 million tons [16, 17]. As the estimate of lithium reserves is difficult, these numbers can fluctuate greatly (Fig. 4).

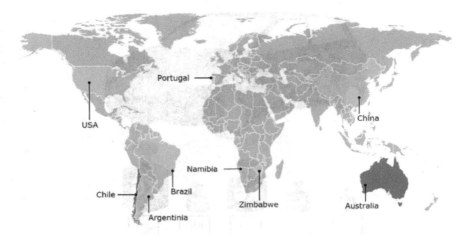

**Fig. 4.** Leading lithium-producing countries according to [13, 15, 21]

Australia is currently the leading lithium supplier with a world market share of over 65% in 2018 [3]. With 18% of lithium-production in 2018, Chile occupies second place behind Australia, and furthermore has the world's largest reserves of the raw material by holding more than 50% in global lithium deposits [3]. Chile's lithium occurs in underground salt lakes. Salt water containing lithium is brought to the surface and to evaporate in large pools, taking advantage of the climatic conditions of the desert. The remaining salt solution is further processed until lithium is suitable for use in batteries.

In Australia lithium is extracted from spodumene ore, which can be processed directly into lithium hydroxide. This gives Australia an advantage over producers who extract lithium from the brine of salt lakes in South America [8].

### 3.2 Environmental and Human Rights Issues

Environmental concerns are being raised especially with regard to lithium mining in the desert regions in South America (Fig. 5). Whether the tapping of the underground lakes has an impact on the groundwater level is controversial. Critics complain that in an already dry environment, natural water resources are being evaporated without being recovered. Therefore, this form of lithium extraction most surely has negative consequences on the ecosystem and the indigenous population. Modern plants with water recovery are a key lever to make lithium production at these sites sustainable for the future.

**Fig. 5.** Salinas Grandes, Andes, Argentina © Kseniya Ragozina | stock.adobe.com

### 3.3 Alternative Sources

Even though Chile and Australia are currently the two main nations for lithium, there are many other lithium mining opportunities across the globe, including Europe. Lithium can be extracted in many places as a by-product in classic mining, so mining can be lucrative even with smaller deposits. Innovative new approaches are also being pursued in Germany. In the Rheingraben in Germany there are large lithium deposits that are to be developed for use in the e-mobility market. A patented approach by Vulcan Energy is coupling lithium production with geothermal systems. The thermal water pumped through the power plant contains large amounts of dissolved lithium. The lithium extracted from the solution in a direct recovery process. Lithium chloride is sent to the refiner and converted to LiOH, the water is recycled. In this way, lithium could be sustainably mined in Germany with the locational advantage due to the physical proximity to many automobile manufacturers. According to the company, lithium production with a negative $CO_2$ balance is possible with this method [23].

## 4 Cobalt

### 4.1 Occurrence and Reserves

Cobalt is thirtieth in the list of elements ranked by frequency, making it a rare element. The world cobalt reserves are estimated at 7.6 million tons [18]. Further cobalt deposits of 120 million tons are suspected in the earth's crust on the floors of the Atlantic, Pacific and Indian Oceans [13]. Unlike lithium, the global distribution of cobalt is very one-sided. Cobalt is mainly obtained from copper and nickel ores. The main ore deposits are in the Democratic Republic of the Congo. Over 70% of the world's cobalt production comes from there [18].

## 4.2 Human Rights Issues and Resulting Trends

Due to the critical mining conditions prevailing in the DRC and the unstable political situation, cobalt can be classified as the ethically most questionable raw material for batteries. Criticism ranges from unsafe working conditions to child labor in illegal mines. This has resulted in special efforts to reduce the use of cobalt in batteries. Future projections assume that global demand for the raw material will increase to 225,000 tons in 2026, of which around 62% will be attributed to battery production. At the same time, the cobalt content in the battery cells has been falling for years, for example due to the fact that the nickel content in the cathode materials NMC and NCA in particular has increased significantly. Companies such as VW and Tesla have also announced battery cathodes instead richer in manganese as a future goal [12, 20].

Cobalt being a very expensive raw material, its amount in batteries should also be reduced for further BEV cost optimization. Both cobalt and lithium prices are highly volatile, with cobalt prices per ton typically being double digit times higher than lithium. This makes recycling particularly interesting.

## 5 Recycling

BEV batteries go through several uses (Fig. 6) – first use as mobile energy storage in the vehicle until its capacity falls below 70–80% of its initial value. Next, the battery serves as a stationary energy storage, for example to buffer temporary energy surpluses from renewable energies such as wind power or photovoltaic systems. The remaining battery capacity per volume and mass plays only a minor role in this application.

Regarding the first use, OEMs most commonly give an 8-year warranty on their traction batteries, expecting real battery life to be well over 10 years. In its second use, the battery will be functional for around another 10 years, before they are eventually materially recycled. Since a large part of the cost is in the battery and its materials (see Sect. 1.2), OEMs are now actively pursuing second life and recycling efforts as part of their business case and have formed alliances with suitable partners [6].

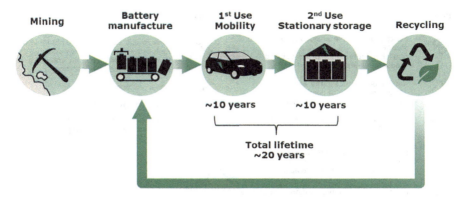

**Fig. 6.** Battery life cycles

Unlike the combustion engine drive, the battery is a closed system, no materials are consumed or lost during its runtime. This makes lithium-ion batteries an excellent candidate for recycling, the ultimate goal being a closed material cycle. Technical studies and pilot projects show that recycling rates of over 90% are possible, VW even aiming at 97% [19]. According to the European Commission's "Circular Economy Action Plan" a recycling efficiency of 65% is to be achieved by 2025 and 70% by 2030. For cobalt, nickel and copper, the target is to increase from 90% in 2025 to 95% in 2030 [7].

## 6 Alternative Battery Technologies

### 6.1 Technical Relevance of Lithium in Batteries

As described in Sect. 2.1, the lithium ions are the charge carriers in the batteries and are therefore the core element for the functioning of the cell. With a standard potential of about −3.05 V, the most negative of all elements, a high cell voltage can be achieved as a result. Furthermore, with lithium being the lightest of all metals and thus leading to the high theoretical capacity of 3.86 Ah/g, lithium has excellent properties for batteries [9].

### 6.2 Beyond Lithium-Ion Technology

In research, efforts are being made to commercialize batteries based on other ions. Sodium-ion technology is in development, Magnesium- and Aluminum-ion batteries are also being researched. Like lithium, sodium is part of the first group of the periodic table and is therefore an alkali metal (see Fig. 2), making it very similar to lithium in its basic properties and is therefore a plausible alternative choice. CATL proved that the use of this technology in automobiles is feasible, when they recently launched a Sodium-ion battery for the EV-market [4].

70   M. Schmalz et al.

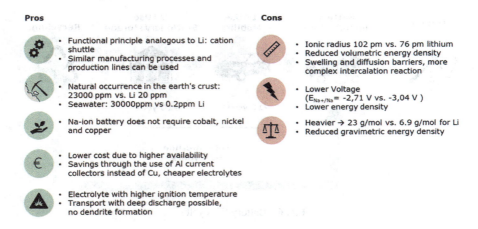

Fig. 7. Pros and cons of sodium-ion batteries in comparison to lithium-ion batteries

In addition to their advantages, the sodium ion technology has major disadvantages, in particular due to the larger ion radius (Fig. 7). Although these are promising approaches for lithium-free batteries, they are largely research topics or niche products whose way towards relevant market share will still be a long one. According to experts, lithium-ion technology will continue to dominate the market for many years to come. Figure 8 shows a timeline of current and expected future battery technologies and materials. It is striking that the vast majority of the next-generation battery technologies, such as the solid state, lithium-sulfur or lithium-air batteries, are still lithium-based. Therefore, lithium will remain the linchpin of battery-powered mobility.

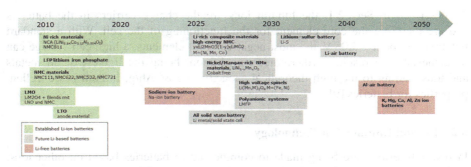

Fig. 8. Current and future battery materials and technologies

### 6.3 Technical Relevance of Cobalt in Batteries

In contrast to lithium, cobalt is merely indirectly involved in the functioning of the lithium-ion battery as an element of the intercalation material. However, it contributes to the thermal stability and thus to the safety of the battery cell, especially in the NMC materials. The result of reduced cobalt levels are often nickel-rich high-energy cells, which entail an increased risk of thermal runaway and propagation. This risk must be controlled by additional measures at the battery module or system level.

### 6.4 Alternatives to Cobalt

As described in Sect. 4.2, a trend towards cobalt reduction in cathode materials has been underway for years. In 2018 Tesla and Panasonic were able to cut the cobalt content in the cathode of the Model 3 batteries to 2.8% compared to the >8% usual at the time [5]. The NMC material currently used in a majority of battery electric cars has also changed significantly in terms of its material content. A few years ago, mixing ratios of the elements nickel, manganese and cobalt in equal parts were common, so-called NMC111. Now nickel-rich variants, the NMC622 or NMC811, have been established and the cobalt content has been gradually reduced.

In addition to the revision of the composition of cobalt-containing cathode materials, there are already several alternative cathode materials that do not require any cobalt at all. As shown in Fig. 1 lithium iron phosphate (LFP) is a viable option. The material, known for being robust, inexpensive and safe, has been on the market for many years but has received less attention due to its lower energy density. Many OEMs such as VW, Tesla or BYD state that they intend to increasingly use LFP cells in the future, especially for small vehicle classes in the volume segment [12, 20].

These developments indicate that according to current forecasts, cobalt is becoming less important in batteries and may disappear altogether. For high-performance vehicles, in which NMC cell chemistries will remain relevant for a longer period of time, recycled cobalt could gain importance.

## 7 Conclusion

Since current as well as many future battery technologies for BEVs are based on lithium, due to its outstanding suitability, the raw material will continue to be of great importance. There are lithium deposits worldwide that can be mined. The goal is to make existing production more environmentally friendly, e.g. through water recycling, and to open up new mining sites all over the world in order to create stable supply chains and prevent temporary delivery bottlenecks and political dependencies.

Contrary to lithium, cobalt is not mandatory for lithium-ion batteries. Viable, well-established alternatives exist and are now gaining in importance. The long-term aim is, to make battery manufacturing a circular economy. A relevant proportion of battery materials will be recycled, technical studies and pilot projects show that recycling rates of over 90% are possible. With an average battery life of around 20 years, however, recycled materials will only gradually make their market entry. In parallel recycling facilities have to experience a large scale-up, that will take place over the next decades.

# References

1. Avicenne Energy: The Rechargeable Battery Market (2017)
2. Breuer, H.: dtv-Atlas Chemie, Bd. 1, 9. Aufl., Deutscher Taschenbuch Verlag (dtv), München, ISBN 3-423-03217-0 (2000)
3. Bundesanstalt für Geowissenschaften und Rohstoffe: Lithium - Rohstoffwirtschaftliche Steckbriefe, Juli (2020)
4. CATL: News, https://www.catl.com/en/news/665.html (03/2022)
5. Electrive.net: https://www.electrive.net/2018/05/31/tesla-model-3-profitabel-durchbruch-bei-kobalt-anteil/ (03/2022)
6. Electrive.net: https://www.electrive.net/tag/second-life/ (03/2022)
7. European Commission, Environment: https://ec.europa.eu/environment/topics/waste-and-recycling/batteries-and-accumulators_en (03/2022)
8. GTAI, https://www.gtai.de/de/trade/australien/branchen/australien-investiert-in-den-abbau-von-batterierohstoffen-617742 (03/2022)
9. https://www.chemie.de/lexikon/Lithium-Batterie.html (03/2022)
10. https://www.regenwald.org/updates/9884/e-autos-von-tesla-gigafabriken-brauchen-gigaminen#fn-kgcenmkh (03/2022)
11. Statista: https://de.statista.com/statistik/daten/studie/159929/umfrage/minenproduktion-von-lithium-nach-laendern/ (03/2022)
12. Tesla Battery Day 2020: youtube.com/watch?app=desktop&v=l6T9xIeZTds (10/2021)
13. United States Geological Survey: Cobalt (2014)
14. United States Geological Survey (2018)
15. United States Geological Survey (2019)
16. United States Geological Survey, 2020: Mineral Commodity Summaries (2020)
17. United States Geological Survey, 2021: Mineral Commodity Summaries (2021)
18. United States Geological Survey (2022)
19. Volkswagen – press release dated 20.02.2019
20. Volkswagen: https://volkswagen-newsroom.com/de/pressemitteilungen/power-day-volkswagen-praesentiert-technology-roadmap-fuer-batterie-und-laden-bis-2030-6891 (10/2021)
21. Volkswagen-AG: News, https://www.volkswagenag.com/de/news/stories/2020/03/lithium-mining-what-you-should-know-about-the-contentious-issue.html (03/2022)
22. Vom Hemdt, A., RWTH Aachen: Batteriezellproduktion und Entwicklung der Wert-schöpfungskette in Deutschland (2019)
23. Vulcan Energy: https://v-er.eu/de/zero-carbon-lithium-de/ (03/2022)

# System Optimization for 800 V e-drive Systems in Automotive Applications

Joao Bonifacio , Felix Prauße, Michael Sperber, Thomas Schupp,
Wolfgang Häge, and Viktor Warth$^{(\boxtimes)}$

ZF Friedrichshafen AG, Friedrichshafen, Germany
{joao.bonifacio,felix.prausse,
michael.sperber,thomas.schupp2,
wolfgang.haege,warth}@zf.com

**Abstract.** Increasing the DC-link voltage up to 800 V brings significant advantages in automotive applications. Faster charging times, lower weight and better efficiencies are some aspects associated with these systems. In this work, a holistic system optimization for an 800 V automotive powertrain with Si-IGBTs and SiC-MOSFETs is discussed. The consideration of the specific challenges for optimizing an inverter for fast switching as well as its consequences at system level are discussed. The choice of the optimal PWM frequency as well as the impact of fast switching on bearing currents are also demonstrated. The efficiency gap between the Si and SiC solutions is derived experimentally and is shown to be very significant in cycle relevant areas. It is demonstrated that only a thorough consideration of the system tradeoffs and physical properties can lead to an optimal solution regarding efficiency, costs and reliability at system level.

**Keywords:** Electrification · Wide-band gap · Automotive applications

## 1 Introduction

Environmental concerns and stringent legislation regarding emissions are having a decisive impact on the adoption of electric vehicles for personal and commercial applications. Despite recent developments in this area, long charging times and limited range are still perceived as drawbacks of this technology by the end-customers. 800 V architectures arise in this context as an answer to these issues. Raising the battery voltage up to these levels enables higher charging powers when compared with current 400 V solutions with the same current. With a higher charging power, it is possible to significantly reduce the charging time or to increase battery capacity, i.e., also increasing vehicle range [1]. Another advantage associated with 800 V architectures are the weight savings across the vehicle especially due to the reduction of the cross section of high voltage cables [2]. Due to the aforementioned

---

© Der/die Autor(en), exklusiv lizenziert an Springer Fachmedien Wiesbaden GmbH, ein Teil von Springer Nature 2023
A. Heintzel (Hrsg.): ATZLive 2022, Proceedings, S. 73–85, 2023.
https://doi.org/10.1007/978-3-658-41435-1_7

reasons, 800 V e-drives have gained much attention in the automotive industry, not only for heavy duty, but also in passenger car applications.

Increasing the battery voltage has, however, several impacts on the e-drive system. At this voltage level, Silicon Carbide (SiC) MOSFETs can offer significant efficiency advantages over the Si-based IGBTs. To maximize this efficiency potential, it is necessary to optimize the whole system design for reaching high voltage steepness, which minimizes the switching losses but also increases EMI issues, voltage overshoots on the semiconductor devices and voltage stress on the insulation system of the electric machine. For reaching the desired voltage steepness without endangering the semiconductors, it is necessary to minimize the DC-link stray inductances. This last aspect brings main design constraints for the power electronics design and must be considered carefully. On the electric machine side, the question regarding the optimal Pulse-Width-Modulation (PWM) frequency, which minimizes the system losses, for the SiC technology and its effects at system level also arises. It becomes clear that a whole system optimization is the necessary solution for designing an 800 V system with maximum performance and efficiency at optimal cost.

In this work, the holistic system development and optimization for an automotive 150 kW 800 V SiC based electric drive is discussed. High voltage steepness up to 40 V/ns is reached and the related switching loss reduction of more than 30 % is demonstrated. Loss reduction at system level compared to the 400 V Si-IGBT system of the same power class reach around 1–2 % in cycle relevant operating regions. The impact of the fast switching SiC power electronics over other system components like bearings is also demonstrated and possible countermeasures are discussed.

## 2  800 V Specific Requirements and System-Level Optimization Issues in Automotive Applications

The design of 800 V e-drive systems must take relevant electrical and mechanical requirements into account. At this voltage level, higher creepage and clearance distances compared to 400 V systems may introduce issues regarding the geometrical placement of the components and the need to develop alternative solutions. These values depend on the application conditions, e.g., altitude, and the voltage levels reached by the system. The definition of the limits for the battery voltage is also paramount for a system optimization. For the system developed in this work, a rated battery voltage of 650 V was set for the power of 150 kW, which was set to be completely available up to the voltage of 820 V. Outside these limits, a power limitation was set and the withstand voltage was defined to be 1150 V. All the components and the geometry of the whole system were designed for coping with these requirements. Figure 1 shows the 800 V inverter developed in this study, which was compatible with both Si-IGBTs and SiC-MOSFETs.

For reaching a cost and efficiency optimal solution at system level, considering the limitations of the inverter, machine and mechanics, it is necessary to analyze in detail the tradeoffs inherent to the system. For example, increasing the switching speed might be an interesting solution for improving the efficiency, but may increase the probability of bearing damage due to parasitic currents and insulation failure. Higher switching speeds also influence the choice of the loss minimal PWM frequency for driving the system, which has also implications regarding costs and computational burden for online control. In general, it can be stated that optimized components do not necessarily yield an optimized system.

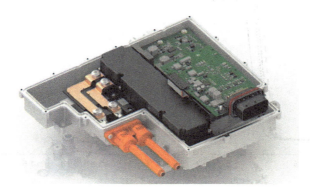

**Fig. 1.** ZF 150 kW 800 V Si/SiC power electronics

## 3 Power Electronics Design for Fast and Efficient Switching

The design of an 800 V SiC power electronics for fast switching is not a straightforward task. Besides the consideration of all normative requirements regarding for example clearance and creepage distances and geometric limitations imposed by the available space, it is also necessary to consider and optimize technical aspects that are influenced by the fast switching. The thermal and supply conditions affect and are also subject to the influence of the switching speed.

It is generally desirable to increase voltage steepness because it yields to a reduction of the switching losses. This reasoning is even more advantageous with SiC because it can reach faster switching due to the high drift velocity of this material [3]. The switching speed can be controlled by optimizing the driver configuration for reaching the desired result. Figure 2 shows drain-source voltages for four different driver configurations at 800 V 465 A. It is possible to see that the reached voltage steepness changes with the driver configurations. Configuration 1 is set to the slowest voltage steepness and Configuration 4 to the fastest one.

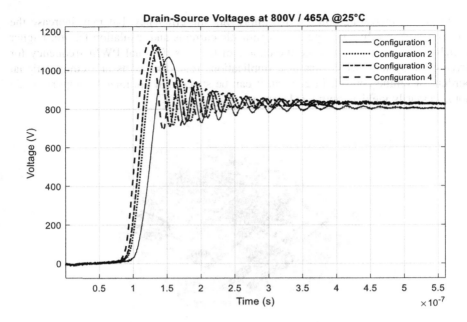

**Fig. 2.** Drain-source turn-off voltages for four different driver configurations at 800 V 465 A

Figure 3 shows the switching energies in per-unit (p. u.) at turn-on and turn-off for the four tested driver configurations. The energy loss reduction between Configurations 1 and 4 reach around 33 and 15 % for the turn-on and turn-off losses respectively. This result highlights the importance and potential of the optimization of the driver configuration for fast and efficient switching.

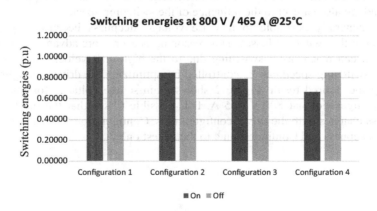

**Fig. 3.** Turn-on and turn-off switching energies (in p. u.) for four different gate-driver configurations

In Fig. 2 it is also possible to see that the drain-source voltage overshoot changes with as a function of the switching speed. In fact, higher switching speeds yield higher voltage overshoots over the switches at turn-off, as it is stated in Eq. (1), where $L\sigma_{DC\_link}$ is the stray inductance of the DC-link. Since the maximum voltage rating for the devices typically used in 800 V 2-level inverters is 1200 V, this limitation must be strictly observed in order to guarantee an operation inside the Safe-Operation-Area. It becomes clear that a minimization of the DC-link's stray inductance is a mandatory step to reach high switching speeds.

$$v = L\sigma_{DC\_link} \frac{di}{dt} \tag{1}$$

The total DC-link stray inductance is defined by the sum of the stray inductances of the power-module ($L\sigma_{PM}$), of the mechanical interconnection between DC-link capacitor and power-modules ($L\sigma_{mech}$) and of the DC-link capacitor itself ($L\sigma_{Cap}$), as shown in Eq. (2).

$$L\sigma_{DC\_link} = L\sigma_{PM} + L\sigma_{mech} + L\sigma_{Cap} \tag{2}$$

For the design of the present system, a very low stray inductance of 9.15 nH has been reached, as demonstrated by the results in Table 1. It is also important to highlight that the simulation and measurement of the stray inductance match very well. The validation of the simulation tools for this aspect of the power electronics was an enabler for performing the optimizations which lead to the final low value of stray inductance This was accomplished by a thorough optimization of the DC-link capacitor design and its connection with the power module. The rationale of these optimization was to reduce to a minimum the size of the busbars between power-module and DC-link capacitor as well as finding a geometric placement of the HV-DC busbars which ensures low stray inductances while providing good cooling properties.

**Table 1.** Simulation and measurement of the DC-link's stray inductance

|                   | Stray inductance |
| ----------------- | ---------------- |
| Simulation [nH]   | 8.9              |
| Measurement [nH]  | 9.15             |
| Deviation         | 2.81 %           |

With these optimizations at driver level and regarding the DC-link's stray inductance, it was possible to reach a voltage steepness of 40 V/ns with the optimized setup. Further optimizations are possible and still ongoing since the parameters of the power-module are also currently being optimized to reach even lower values of DC-link stray inductances and higher switching speeds.

Another important requirement for an 800 V SiC power electronics, is that it should allow higher PWM frequencies for minimizing system losses and DC-link ripple. The effect of the PWM frequency at system level will be discussed in the following section. For the inverter design, increasing the switching frequency has dramatic consequences for its thermal design and impacts the choice of components that can be used in a particular application. Figure 4 (left) show the supply current needed by the driver board as a function of the switching frequency. The supply current increases, as expected, in order to cope with increased switching energy needed for higher PWM-frequencies. The current increase defined by the targeted frequency will define the design of the power supply circuits as well as the components that will be selected for driving the power-modules. In this specific case, the maximum frequency was chosen to be 30 kHz.

In Fig. 4 (right) it is also possible to see the implementation of the connection between driver board and the power-modules. The losses on the driver board increase with increasing current for driving the switches at higher frequencies. Due to this reason, a Thermal Interface Material (TIM) and a power-module bracket are used for cooling the driver-board. This effect is reached by thermally connecting the driver circuit with the cooling channels below the power-modules. Simulations and measurements show that a temperature reduction of up to 40 °C can be reached with such configuration.

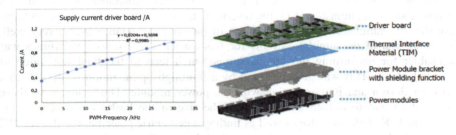

**Fig. 4.** Necessary supply current for the driver board (left) and cooling concept of power-modules and driver board using a Thermal Interface Material (right)

Another function achieved by the power-module bracket is shielding the driver board from the electromagnetic noise produced by the power-modules. This function becomes more important for fast-switching converters, since the risk of electromagnetic interference increases significantly.

## 4 The Optimal Choice of PWM Frequency

As demonstrated in the previous sections, tuning the switching speed of 800 V SiC power electronic devices enables a significant loss reduction under operation. Another important parameter, which also has an important impact on the losses at system level is the choice of the PWM-frequency. It is known from literature [4, 5] that the high-frequency electric machine losses reduce asymptotically with increased switching frequency. At the power electronics, the switching losses increase linearly with the PWM frequency [6]. Therefore, there is a unique switching frequency which minimizes the overall losses at system level. In modern automotive e-drives a full variable switching frequency strategy is typically implemented in order to guarantee an energy optimal state at every operating point.

Figure 5 shows the loss difference between the frequencies of 10 and 4 kHz for the electric machine and the 800 V SiC inverter at two different operating points: 5000 rpm with 100 Nm and 4250 rpm and −275 Nm. The variation of the electric machine losses is mainly negative – except for some outliers which were caused by measurement and interpolation errors at border regions, whereas the variation of inverter losses is mainly positive. This confirms the expectation that the e-machine losses decrease while the inverter losses increase with higher PWM frequencies. It is also possible to conclude that the potential for loss reduction at system level through a manipulation of the PWM-frequency depends strongly on the operation point. Figure 6 shows the loss difference for machine and inverter for 16 and 10 kHz. The machine losses, although still negative, are lower compared to the variation between 4 and 10 kHz, which indicates that the potential for high frequency loss reduction of the machine reduces with increasing PWM frequency. For the inverter, the losses are positive as in the previous case.

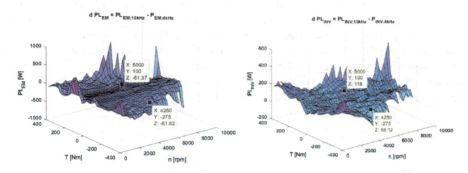

**Fig. 5.** Loss difference between PWM frequencies of 10 and 4 kHz for e-machine (left) and inverter (right)

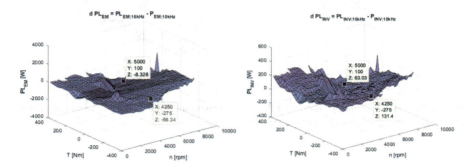

**Fig. 6.** Loss difference between PWM frequencies of 16 and 10 kHz for e-machine (left) and inverter (right)

Measurements up to 28 kHz have been carried out with the SiC system for establishing the optimal switching frequency for each operating point. Figure 7 (left) shows the results of the optimization using only the overall loss at system level as criteria. The optimal PWM frequency lies around 10 kHz in the low torque region and around 14 kHz at the high torque region. Frequencies higher than 14 kHz lead to increased losses at system level because the higher switching losses at the inverter are not fully compensated by a loss reduction at the e-machine side. For the optimization two boundary conditions were also considered: the minimal switching frequency for ensuring the controllability of the electrical machine, which is particularly relevant at higher speeds and the DC-ripple boundary, which might prevent the use of very low switching frequencies in order to keep the DC-ripple in an acceptable range. In Fig. 7 (right) it is possible to see that for the SiC system the controllability and the DC-ripple conditions were always satisfied with the efficiency optimized frequency. This is the case because the optimal frequencies are relatively high, due to the reduced weight of the switching losses compared to the high frequency losses of the machine.

Performing the same optimization with a Si power electronics yields the results shown in Fig. 8. The optimal frequencies are rather low compared to the optimal ones from the SiC system, ranging from 2 to 6 kHz. If the boundary conditions regarding the controllability and the DC-ripple are considered, it is possible to see in Fig. 8 (right) that the controllability condition determines the PWM frequency at higher speeds and the DC-ripple condition determines the frequency for some points at the mid-speed range. This can be explained by the fact that the efficiency optimal frequencies for Si are rather low because the switching losses of the inverter are more significant compared to the e-machine high frequency losses. It is important to mention that other aspects such as NVH (noise, vibration and harshness) will also have an impact on the choice of the optimal PWM frequency.

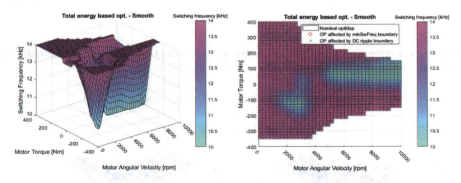

**Fig. 7.** Operation point dependency of the optimal switching frequency (left) and consideration of boundary conditions (right) for the 800 V SiC system

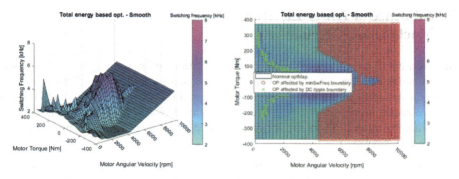

**Fig. 8.** Operation point dependency of the optimal switching frequency (left) and consideration of boundary conditions (right) for the 800 V Si system

It is possible to conclude from these energy optimizations that the use of SiC enables and requires the application of PWM-frequencies that are higher than the ones typically used for Si-IGBTs. For assessing the loss reduction potential of 800 V SiC compared to the 400 V Si solution, measurements at rated voltage were carried out for both systems. In Fig. 9 the loss difference between both systems is shown. The 400 V-Si system was operated with 325 V voltage and 8 kHz switching frequency whereas the 800 V -SiC system was operated at 650 V and 10 kHz. The electric machine was the same in both cases, with the only difference being the windings connection for allowing the operation at higher voltages. The measured loss difference reaches up to 2–3 %, indicating the 800 V SiC system is more efficient than the 400 V-Si. In cycle relevant areas, the loss advantage that can be achieved lies around 1–2 %. This result confirms the expectation that the use of SiC has a significant potential for increasing the efficiency and consequently the range of electric vehicles [7].

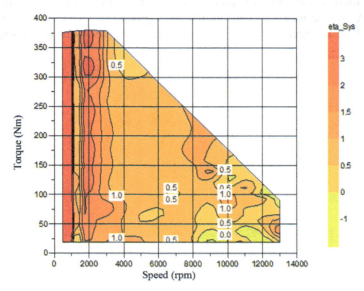

**Fig. 9.** Loss comparison at system level between Si-IGBTs at 325 V 8 kHz and SiC-MOSFETs at 650 V 10 kHz with an automotive PMSM

## 5 Influence of Power Electronics Parameters on the Electric Machine and Mechanics

An often-neglected aspect of a whole e-drive optimization is the influence of power electronics parameters on other components, especially on the mechanics' side. One of the most significant aspect of these interactions is the effect called bearing currents. These currents typically arise in power-electronics driven e-drive systems because of the switching events. There are in general three types of power-electronics induced bearing currents [8]:

- EDM (electric discharge machining) currents: These currents arise as a pulse when the voltages over the bearings exceed the threshold voltage of its lubrication film. The voltage over the bearing mirrors the common-mode voltage and is proportional to a ratio between electric machine and bearing capacitances (the so-called bearing voltage ratio, which is geometry dependent).
- Circulating bearing currents: The pulsating voltages in the machine terminals cause a ground current at the stator. This current excites a circular magnetic flux, which induces a voltage along the shaft of the e-machine. If this voltage is high enough to break the insulation film, a current flowing from one bearing to the opposite one will appear.
- Rotor ground currents: These currents appear when the machine is grounded by the load and can reach considerable magnitudes.

Bearing currents are a very critical phenomena because they generate premature failure of the bearings, reducing the useful life of the whole e-drive system. In this project, a measuring setup was designed for characterizing the bearing currents. Figure 10 shows the e-drive shaft, which contains three bearings (L1, L2 and L3, from left to right). For measuring the bearing currents, a set of five Rogowski-coils has been used. These coils are enumerated from L1_1 to L3 and placed as shown in Fig. 10. A voltage measurement through shaft grounding elements has been placed in L1 and L2 in order to measure all relevant characteristics of the arising bearing currents.

System Optimization for 800 V e-drive ... 83

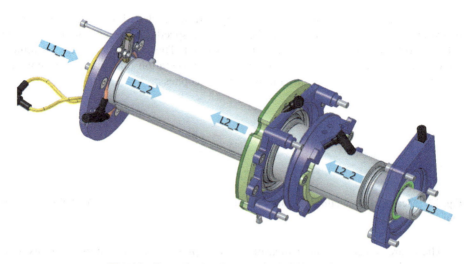

**Fig. 10.** Setup for bearing currents' characterization

Figure 11 shows the bearing currents' measurement at 3000 rpm, 0 Nm and 650 V with Si-IGBT switching devices. The currents measured by the Rogowski-coils are equal in all measuring points, which indicates that circular currents are present in this e-drive. At each switching event, a current flows into the shaft from the left side of bearing L1 and finally goes back to the housing through bearing L3. The current could also in some cases flow through the gears between L2 and L3 into other bearings in this path. No EDM currents have been detected in this system. It is important to highlight that these measurements have been carried out without any counter-measure for better assessing and characterizing the bearing currents.

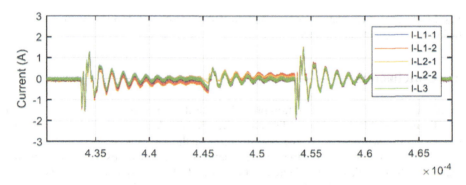

**Fig. 11.** Circulating currents at 3000 rpm, 0 Nm and 650 V with Si-IGBTs

In Fig. 12 a comparison between bearing currents under Si and SiC at 5000 rpm, 0 Nm and 800 V can be analyzed. The switching speeds are around 40 V/ns for the SiC-MOSFETs and 2 V/ns for the Si-IGBT variant. The current under SiC is almost the double of that under Si, which confirms the expectation that circular currents are negatively affected by increasing the switching steepness.

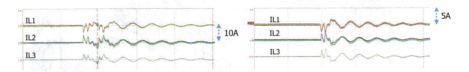

**Fig. 12.** Measured circulating currents with SiC (left) and Si (right) at 800 V, 0 Nm and 5000 rpm

There are several counter-measures for coping with circulating currents that can be applied to an automotive e-drive system. The selection process for the most suitable strategy in a particular case must take several aspects into account, e.g. available installation space, durability requirements, application environment and costs. Figure 13 shows the effectiveness of an AC filter for reducing the circulating currents at 800 V, 0 Nm and 5000 rpm.

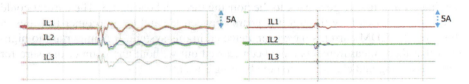

**Fig. 13.** Measured circulating currents with Si-IGBTs at 800 V, 0 Nm and 5000 rpm without (left) and with (right) an AC-filter

## 6 Conclusions

In this work, a holistic optimization procedure for designing an 800 V system with Si-IGBTs and SiC-MOSFETs was discussed. Besides the design constraints for reaching fast switching with SiC, like the optimization of the DC-link stray inductances and the component choices for the driver board, the implications of fast switching regarding the overall system efficiency, the optimal PWM frequency and EMC effects have been discussed. The efficiency advantages of SiC related to Si became very clear from the measurement of the overall system losses, as well as the possibility offered by SiC of using higher PWM-frequencies for minimizing the system losses. The overall efficiency advantage of 800 V -SiC compared to 400 V-Si lies around 1–2 % in cycle relevant operating regions. Moreover, the effects of the fast switching on the properties of bearing currents, as well as the effectiveness of an AC-filter for coping with this issue, have been demonstrated experimentally. These issues indicate that only a holistic and interdisciplinary design approach can yield a functional, cost competitive and efficient automotive e-drive system.

## References

1. Aghabali, I., Baumann, J., Kollmeyer, P., Wang, Y., Bilgin, B., Emadi, A.: 800-V electric vehicle powertrains: review and analysis of benefits, challenges and future trends. IEEE Trans. Transp. Electr. **7**(3) (2021)
2. Jung, C.: Power up with 800-V systems: the benefits of upgrading voltage power for battery-electric passenger vehicles. IEEE Electr. Mag. **1**(5) (2017)
3. Shang F., Arribas, A.P., Krishnamurthy, M.: A comprehensive evaluation of SiC devices in traction applications. IEEE Transp. Electr. Conf. Expo (ITEC) (2014)
4. Rasilo, P., Martinez, W., Fujisaki, K., Kyyrä, J., Ruderman, A.: Simulink model for PWM-supplied laminated magnetic cores including hysteresis, eddy-current, and excess losses. IEEE Trans. Power Electron. **34**(2) (2019)
5. Zhang, D., Liu, T., Zhao, H., Wu, T.: An analytical iron loss calculation model of inverter-fed induction motors considering supply and slot harmonics. IEEE Trans. Ind. Electron. **66**(12) (2019)
6. Stempfle, M., Fischer, M., Roth-Stielow, J.: Loss modelling to optimize the overall drive train efficiency. In: 17th European Conference on Power Electronics and Application – EPE'15 ECCE-Europe
7. Hain, S., Meiler, M., Denk, M.: Evaluation of 800 V traction inverter with SiC-MOSFET versus Si-IGBT power semiconductor technology. In: Proceedings of PCIM Europe, Nuremberg, Germany (2019)
8. Plazenet, T., Boileau, T., Caironi, C., Nahid-Mobarakeh, B.: A comprehensive study on shaft voltages and bearing currents in rotating machines. IEEE Trans. Indus. Appl. **54**(4) (2018)

# FCTRAC and BioH$_2$Modul – A Way to Zero Emission Mobility in Agriculture

Veronica Gubin[1]([⊠]), Christian Varlese[2], Florian Benedikt[1],
Johannes Konrad[2], Stefan Müller[1], Daniel Cenk Rosenfeld[1],
and Peter Hofmann[2]

[1] Institute of Chemical, Environmental and Bioscience Engineering,
Technische Universität, Wien, Austria
{veronica.gubin,florian.benedikt,stefan.mueller,
daniel.rosenfeld}@tuwien.ac.at
[2] Institute for Powertrains and Automotive Technology,
Technische Universität, Wien, Austria
{christian.varlese,johannes.konrad,
peter.hofmann}@ifa.tuwien.ac.at

**Abstract.** On the way to Zero Emission Mobility, no feasible solution for tractors is so far available, as battery electric vehicles are not able to fulfill the requirements for range, refueling time, and weight. Therefore, in the frame of the funded project "FCTRAC", a fuel cell tractor is developed from an existing diesel vehicle to meet these specific requirements of the powertrain. Due to the limited space in the vehicle and the challenges arising from the thermal management of the fuel cell powertrain, innovative solutions have been developed to achieve the full-operability of the vehicle.

Moreover, no hydrogen fueling infrastructure exists in the operational areas of most of tractor use-cases. Hence, another goal of the project is the development of an input-flexible plant for decentralized production of green hydrogen. Product gas from gasification of wood chips, biogas out of biogas plants, as well as digester gas out of sewage treatment plants are considered as input gases. The proposed solution, the so-called "BioH$_2$Modul", can be coupled with these sources and deliver hydrogen to the storage system. As combined heat and power plants using wood gasification, sewage treatment plants, biogas plants are widespread in the agricultural sector, hydrogen can be produced in the operational areas of the tractors.

The first part of the paper will present the wide range of process units, which can be implemented in the process chain design for decentralized production of high-purity hydrogen from the different feed gases. These process units are consequently classified according to their operating conditions, thus forming the basis for the design of the process chains.

In the second part, the fuel cell powertrain will be described as well as the thermal system and simulation results. Through an innovative packaging in the vehicle, the higher cooling requirements of the fuel cell system are fulfilled even in critical conditions.

---

© Der/die Autor(en), exklusiv lizenziert an Springer Fachmedien Wiesbaden GmbH,
ein Teil von Springer Nature 2023
A. Heintzel (Hrsg.): ATZLive 2022, Proceedings, S. 86–104, 2023.
https://doi.org/10.1007/978-3-658-41435-1_8

**Keywords:** Fuel Cell Tractor · Green Hydrogen Production · Thermal Management

# 1 Project Introduction

## 1.1 FCTRAC Goals

The project "FCTRAC" addresses the challenges of developing a fuel cell tractor (FCTRAC) and the local production of green hydrogen ($H_2$) in a symbiotic relationship with its ecosystem. In the frame of this project, a broader vision has been adopted, embracing the vehicle and its operating environment as an interconnected system. Whereas carbon-neutral solutions for passenger cars and trucks are already available on the market, tractors have received marginal attention in this decarbonization process despite their wide field of applications. Moreover, in the recent debate over sustainable mobility, a common misconception has arisen from neglecting the production of carbon-neutral primary energy sources, as the focus has shifted towards the mere development of locally zero-emission vehicles regardless of the impact of emissions from energy sources. The approach of FCTRAC goes beyond these misconceptions and provides a viable solution for decarbonization, where the needs of our society become sustainable and in harmony with our planet's resources.

$H_2$ is produced on-site from sustainable biomasses in rural areas, which usually are the operation area of tractors. In order to achieve this closed cycle between production and utilization, the so-called "BioH$_2$Modul" is under development. Within the BioH$_2$Modul green $H_2$ can be produced from different feed gases, such as product gas (PG) from wood gasification, biogas (BG) out of biogas plants, and digester gas (DG) out of sewage treatment plants. Along with the refueling infrastructure, an existing tractor powered by diesel is going to be retrofitted with a fuel cell powertrain (FC-PT) and be deployed directly where the residue feedstock is derived.

This project is funded by the Climate and Energy Fund, the Institute for Powertrains and Automotive Technology of TU Wien has the role of project coordinator and several partners of Austrian industry are involved: AVL List GmbH, CNH Industrial Österreich GmbH, Engineering Center Steyr GmbH & Co KG, Glock Technology GmbH, HyCentA Research GmbH, Sohatex GmbH and TU Wien Institute of Chemical, Environmental and Bioscience Engineering.

# 2 Input-flexible Renewable Hydrogen Production from Different Biomass Sources

Within the scope of the project "FCTRAC", a plant for decentralized $H_2$ production, the so-called "BioH$_2$Modul", is designed, built and operated in order to produce high-purity $H_2$ for polymer electrolyte membrane (PEM) fuel cell (FC) application. Input-flexibility and thus modularity of the process chain are part of the plant concept, so that other feed gases could also be converted into high-purity $H_2$ after restructuring the process chain.

Different gas sources derived from biomass and residues, that are available in rural areas in Austria, are considered with regard to decentralized $H_2$ production. (a) PG from gasification of wood chips, (b) BG out of biogas plants, as well as (c) DG out of sewage treatment plants are therefore from great interest as feed gases. However, only PG from air gasification of wood chips, which is usually converted into heat and power within a combined heat and power (CHP) plant, is considered as design basis of the "BioH$_2$Modul" for the demonstration phase. Due to additional $H_2$ production for FC tractors in low-energy seasons, the overall efficiency of CHP plants is improved as constant plant utilization is achieved. In Fig. 1, the scheme of sustainable $H_2$ production for FC application in a FC tractor can be seen.

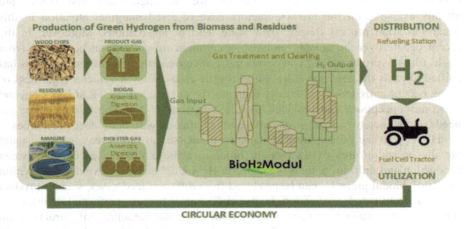

**Fig. 1.** Concept of sustainable and input-flexible hydrogen production from biomass for fuel cell tractor application aiming at a circular economy approach

### 2.1 Biomass as Original Feedstock

Modern bioenergy is becoming increasingly important for the transition of the energy system towards a high share of renewable energy sources. The World Energy Outlook 2021 clearly shows that modern bioenergy will play a growing role in all three evaluated scenarios, including the Net Zero Emission Scenario (NZE). The NZE scenario follows a narrow pathway to achieve net zero carbon dioxide ($CO_2$) emissions in the global energy sector by 2050. In addition, the conventional use of biomass is no longer considered in the NZE Scenario [1].

Modern bioenergy comprises conversion technologies with solid, liquid and gaseous products as secondary energy carriers [2]. In order to obtain a gaseous renewable energy carrier as feedstock for $H_2$ production, the solid biomass first has to be converted either thermochemically, by pyrolysis, whereby pyrolysis is here not further considered, or gasification, or biochemically, by anaerobic digestion [3].

In the gasification process, solid biomass is converted into PG using different gasification agents in dependence of the heat supply concept. Whereas in case of auto-thermal gasification, the necessary heat is produced by partial oxidation of the bio-mass itself, allothermal gasification requires external heat input by heat exchanging or circulating bed material [4]. In this project, PG both from air gasification within a fixed-bed downdraft (FIXB) gasifier, representing an autothermal process, and from steam gasification within a dual fluidized bed (DFB) gasifier, an example for an allothermal process, is being considered. For further detailed information on the gasification technologies see [4].

During anaerobic digestion, biomass is transformed by micro-organisms in the absence of oxygen, depending on the feedstock, into BG or DG. While in biogas plants, non-lignocellulosic biomass with high water contents is utilized, in sewage treatment plants, sewage sludge is converted into DG for sludge stabilization and volume reduction [3, 4].

## 2.2 Biomass-based Feed Gases

As gases with widely varying properties are produced by gasification and anaerobic digestion, the process chain for $H_2$ production has to be appropriately designed. In the concept of the "BioH$_2$Modul", the main gas input composition essentially determines which process units might be used, whereby their arrangement might be made according to concentrations of impurities as well as feed gas temperature, pressure, and mass flow.

Compositions of FIXB-PG and DFB-PG from gasification as well as BG and DG are shown in Table 1. In contrast to BG and DG, FIXB-PG and DFB-PG are

**Table 1.** Compositions of product gases from air and steam gasification as well as biogas and digester gas

| Gas composition | Unit | FIXB-PG [7] | DFB-PG [8] | BG[1] [9] | DG [9] |
|---|---|---|---|---|---|
| $H_2$ | vol.-%$_{db}$ | 17.2 | 43.8 | n. a | n. a |
| CO | vol.-%$_{db}$ | 21.2 | 22.3 | n. a | n. a |
| $CO_2$ | vol.-%$_{db}$ | 12.6 | 20.2 | 45.9/36.1 | 35.6 |
| $CH_4$ | vol.-%$_{db}$ | 2.5 | 9.7 | 53.1/62.2 | 64.2 |
| $N_2$ | vol.-%$_{db}$ | 46.0 | n. a | 1.2/0.5 | 0.2 |
| $O_2$ | vol.-%$_{db}$ | n. a[3] | n. a | 0.2/0.1 | 0.1 |
| $C_xH_y$[3] | vol.-%$_{db}$ | 0.4 | 1.0 | n. a | n. a |
| $H_2O$ | vol.-% | n. a | 36.0 | [5] | [6] |

---

[1] Left: BG from renewable raw materials; Right: BG from residues; Average values from different measurements do not necessarily result in 100 %.

[2] n. a. – not available.

[3] $C_xH_y$ – light hydrocarbons such as ethane, ethylene, propane.

[4] Saturated gas according to temperature and pressure in digester.

[5] "

already $H_2$-rich gases with a carbon monoxide (CO) content of approximately 21–22 vol.-%$_{db}$[6], whereby the former gases contain more than 50 vol.-%$_{db}$ methane ($CH_4$) and have significantly higher $CO_2$ contents. Due to air as gasification agent, the nitrogen ($N_2$) content in FIXB-PG is quite high in comparison to the other gases. However, on account of steam as gasification agent, DFB-PG contains considerably more water ($H_2O$). PG also contains impurities such as particulate matter (PM), tar, alkalis, ammonia ($NH_3$), hydrogen sulfide ($H_2S$), other sulphur compounds, and hydrogen chloride (HCl) [4]. In BG and DG, trace gas components such as $N_2$, oxygen ($O_2$), CO, $H_2S$, other sulphur compounds, $NH_3$, halides, and siloxanes can be found [5, 6]. Air quite often gets into the digester, which is indicated by the $O_2$ and $N_2$ content [5]. The FIXB-PG from [7] was obtained by air gasification of wood sawdust pellets.

## 2.3 Process Units for Conversion Technologies and Hydrogen Purification

In the following sections, commercially applied process units for conversion of $CH_4$ and CO, removal of $CO_2$ and $H_2O$ as well as $H_2$ purification are outlined.

**Methane and Light Hydrocarbons.** In the methane reforming process, $H_2$ and CO are obtained by adding steam and/or oxygen respectively air to $CH_4$. If steam is used, the process is called steam methane reforming (SMR). SMR is carried out in a temperature range of 500–900 °C in the presence of a Ni-based catalyst. Although, low pressure is thermodynamically favorable for SMR, industrial processes operate at elevated pressures as $H_2$ is commonly used for methanol or ammonia synthesis. However, in the case that steam and oxygen respectively air is used, autothermal reforming (ATR) occurs. ATR processes are operated at 900–1150 °C, usually applying two stage catalysts. The addition of air respectively oxygen leads to partial oxidation (POX two-stage) of methane at a temperature rage of 1150–1500 °C. Besides methane, light hydrocarbons ($C_xH_y$) are also reformed in these processes [10]. Sulphur compounds have to be removed previously to avoid poisoning of the catalysts. Within the SMR process, a relatively high $H_2$/CO ratio can be obtained compared to ATR and POX. Therefore, SMR is the preferred reforming process for $H_2$ production [10, 11].

**Carbon Monoxide.** As even low CO contents cause irreversible damage to the performance and life time of FC [12], lowering the CO content to a ppm level is essential. In industrial $H_2$ production processes, the water-gas shift (WGS) reaction is applied usually as two-stage reaction system. Within these reactors, CO is converted into $CO_2$ and $H_2$ by adding steam in the presence of catalysts. The slightly exothermic and reversible reaction is desired to run at low temperatures to achieve low CO contents without adding a large amount of excess steam. Pressure does not significantly affect the equilibrium. While in the first stage, the high temperature

---

[6] Db – dry basis.

shift (HTS) stage, Fr/Cr-based catalysts are used at inlet temperatures in the range of 350–550 °C, the low temperature shift (LTS) stage, is equipped with a Cu/Zn-based or CoMo-based catalyst operating at an inlet temperature level of 250 °C. The Cu/Zn-based catalyst is not tolerant to sulphur contents above 0.1 ppm. However, Co/Mo-based catalysts are sulphur-resistant and require a certain ratio of $H_2S$ to $H_2O$ in the feed gas after catalyst pretreatment to preserve the active form [10].

**Hydrogen Purification.** Membrane separation and pressure swing adsorption (PSA) are commercially established processes for $H_2$ separation and purification. Although the membrane separation process is normally pressure-driven in gas separation applications [10], the driving forces can also include concentration and potential difference in addition to the pressure difference [12]. Both processes provide a continuous $H_2$ flow, whereby the PSA system is cyclically operated and therefore requires more complex process control than membrane separation. In membrane modules, $H_2$ selectively permeates the membrane while all the remaining impurities such as CO, $N_2$, $C_xH_y$, $CO_2$, or $CH_4$ exit the membrane module as they have lower relative rates of permeation [10, 13]. In contrast to membrane separation, $H_2$ is produced in the PSA process at approximately feed pressure. Impurities are removed from the feed gas by selective adsorption in fixed-bed columns consisting of activated carbon (AC), zeolite, silica or alumina gel. By lowering the partial pressure in the column, the adsorbed impurities CO, $CO_2$, $N_2$, $CH_4$, and $C_xH_y$ are desorbed from the adsorbent bed [10]. For this reason, the PSA system contains multiple columns to provide a continuous flow of $H_2$. The selection of appropriate adsorbents for $H_2$ purification depends on the thermodynamic selectivity for different impurities over $H_2$ and desired fast adsorption kinetics [10]. Cryogenic distillation is used for large-scale $H_2$ purification at low temperatures and elevated pressure by exploiting its high volatility in comparison to other gas components such as $CH_4$, $C_xH_y$, CO, and $N_2$. However, $CO_2$, $H_2S$, and $H_2O$ have to be removed first [12, 14].

**Carbon Dioxide.** $CO_2$ removal technologies include both chemical and physical absorption, PSA and membrane separation. While physical absorption is accomplished by water or organic physical scrubbing, chemical absorption is carried out by amine scrubbing [15]. $CO_2$ can be separated from the feed gas with water, whereby both media are pressurized. In case of biogas upgrading, the biogas pressure usually amounts to 1000–2000 kPa. Simultaneously, $H_2S$ and $NH_3$ are also removed. Physical organic scrubbing, in contrast, uses polyethylene glycol or methanol. These processes are operated at high pressure and low temperature [5, 10, 16]. Amine scrubbing is applied for chemical absorption of both $CO_2$ and $H_2S$ at a temperature level of approximately 25–60 °C [10].

**Water.** $H_2O$ removal technologies comprise, on the one hand, physical drying methods such as condensation, and, on the other hand, chemical drying methods. The latter method includes adsorption drying on silica, alumina, or molecular sieves as well as absorption with polyethylene glycol or hygroscopic salts. The most common technology for chemical water drying in case of BG upgrading is adsorption on

92        V. Gubin et al.

alumina or molecular sieves (zeolites). Polyethylene glycol scrubbing might also separate dust, organic sulphur compounds, $H_2S$, and $NH_3$. Chemical methods are usually operated at higher pressure levels [5].

### 2.4  Process Units for Trace Contaminants Removal

In this chapter the following trace contaminants are discussed: PM, tars, $H_2S$, other sulphur compounds, $NH_3$, halides, and siloxanes.

**Particulate Matter.** The most relevant forms of PM within this study are dust and fly char. Depending on the particle content and size as well as requirements of downstream processes, cyclones, filters with a filter medium, electrostatic precipitators (ESP) or scrubbers can be used for removal of PM. As cyclones can be applied over a wide temperature range, they are suitable for removal of PM at hot temperatures. Whereas fabric filters are used at gas temperatures below 250 °C, rigid filters can be operated at temperatures below 900 °C, as the filter candles are usually made of metal or ceramic. With ESP, both dry and wet separation can be carried out, whereby the use of a wet electrostatic precipitator (wESP) is limited to a temperature level of 65 °C. PM is also removed in water and biodiesel scrubbers [4].

**Tars.** *"The organics produced under thermal or partial oxidation regimes (gasification) of any organic material are called "tars" and are generally assumed to be largely aromatic"* [17]. Physical tar removal processes are widely used and include dry technologies such as cyclones or filters and wet technologies such as wESP and biodiesel scrubbing [4, 18]. Tar removal is also accomplished in-situ with primary measures in the gasifier by thermal and catalytic cracking [16]. AC fixed-bed filter can alternatively be used for dry tar removal [17, 19, 20].

**Hydrogen Sulfide and other Sulphur Compounds.** Sulphur compounds are commonly removed by wet as well as dry desulphurization, and by catalytic adsorption. Wet desulphurization comprises absorption technologies using solvents for either chemical or physical absorption. Physical absorption becomes attractive when the gas is available at high pressure. In case of amine scrubbing, representing chemical absorption, and cold methanol scrubbing, representing physical absorption, the sulphur content can be lowered down to 0.1 ppm. For dry desulfurization, various adsorbents are applied, such as metal oxides, AC or zeolites. Zinc oxide, for example, is used to lower the $H_2S$ content to ppb at medium operating temperatures in the range of 300–400 °C. However, desulphurization with AC or zeolite is carried out at ambient operating temperature. By catalytic hydrodesulfurization (HDS) at 200–300 °C, other sulphur compounds can be converted into $H_2S$ by adding $H_2$. Subsequently, $H_2S$ can be removed either by wet or dry desulphurization methods [10].

**Ammonia and other Nitrogen Compounds.** During gasification, most of the nitrogen in the biomass is converted into $NH_3$ and a lesser extent to hydrogen cyanide

(HCN) [4]. Ammonia removal is carried out by adsorption on AC or water scrubbing [5, 16].

**HCl and other Halides.** HCl is the most common halide derived from biomass gasification. Commercially, HCl is removed in cold gas cleaning stages along with alkali, tar, and PM or by hot gas cleaning technologies such as adsorption on AC, alumina, or alkali. HCl can be scrubbed by water, as HCl is highly soluble in water [16]. AC can be applied for adsorption of HCl also at ambient temperature [21]. Halides, such as halogenated hydrocarbons, can be found after anaerobic digestion due to volatilization of halogen-containing materials and are removed by adsorption on AC [22].

**Siloxanes.** Siloxanes might be trace compounds in DG, as they are ingredients of deodorants as well as shampoos and can thus also be found in sewage sludge [15]. These impurities are commercially removed either by adsorption technologies or in deep chillers. The formers include fixed-bed adsorption on AC as well as fixed-bed bed temperature swing adsorption (TSA) and fluidized bed adsorption [6]. However, it is reported in [5] and [23] that AC adsorption of siloxanes is the most common method among the available technologies which is why only AC adsorption is considered in the following. AC adsorbents exhibit varying adsorption capacities for different siloxane molecules at ambient temperature and ambient pressure [24]. To achieve a higher adsorption capacity, a high pressure is required [5].

## 2.5 Selected Process Units in the BioH$_2$Modul

In Fig. 2, selected process units for gas conversion, cleaning, and separation with regard to impurity concentrations, differentiated according to main gas components and trace contaminats, differentiated according to main gas components and trace contaminats, differentiation between main gas components and trace contaminants, applicable for H$_2$ production from different biomass-derived gases can be seen. The process units are classified according to operating conditions, which are limited to the main process without possible regeneration. It is assumed that the siloxane and halide concentrations in BG are significantly lower than in DG and are therefore not included in Fig. 2 as a trace contaminant in BG. As alkali compounds condense and agglomerate in particles or combine with tars at temperature reduction [16], they are not considered in Fig. 2 as further trace contaminant for PG.

Based on this approach, process chains are to be designed for the input-flexible H$_2$ production plant, the "BioH$_2$Modul", with particular attention to temperature, pressure, and water content profiles over the process chains for the different feed gases. For example, repeated gas compression or heating should be avoided in order to prevent the installation of additional compressors or heat exchangers to target an exergetically optimized overall process. Currently, the detail engineering for the "BioH$_2$Modul" is being carried out. The "BioH$_2$Modul" is designed for a production capacity of 3 kg H$_2$ per hour to supply the FC tractor in real operation. Experimental investigations on H$_2$ production from FIXB-PG will be conducted as soon as the plant is commissioned and steady-state operation can be achieved.

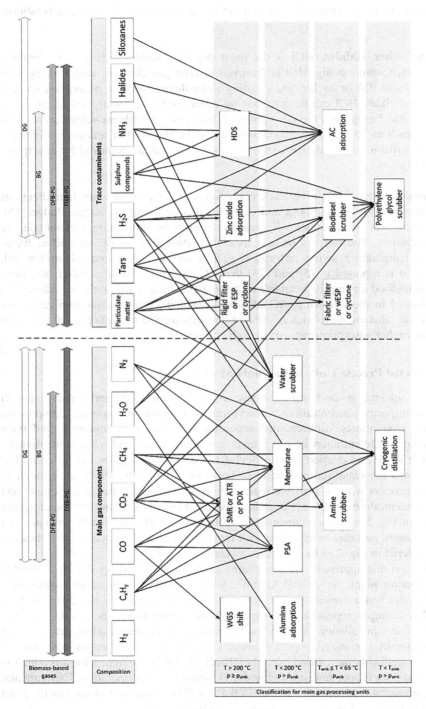

**Fig. 2.** Selected commercially available process units for gas cleaning and gas separation applicable for hydrogen production from biomass-derived gases

## 3 FCTRAC

### 3.1 Vehicle

An existing tractor powered by a 95-kW-diesel powertrain is going to be retrofitted with a FC module of 100-kW net power. A FC-PT was chosen because FC-PTs are superior to lithium-ion battery electric vehicles in terms of life-cycle cost and in particular for the biomass utilization efficiency, as shown in the well-to-wheel analysis of [25]: if the full cycle energy efficiency in a 480-km-driving range is considered, converting biomass to $H_2$ for a fuel cell vehicle uses 40 % less energy than converting biomass to electricity for a battery electric vehicle.

The FC-PT system is shown in Fig. 3 and has been designed to deliver comparable characteristics like the donor vehicle in terms of power output and response. Most of the FC-PT components are off-the-shelf components that were selected for their compactness and reliability. Around 12.4 kg of $H_2$ are stored in the compressed $H_2$ storage system (CHSS) at 70 MPa.

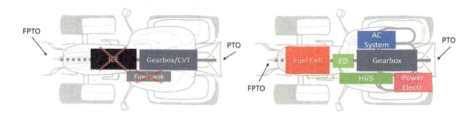

**Fig. 3.** FCTRAC vehicle schematic. Left: donor vehicle. Right: fuel cell tractor. (FPTO: front power take-off, ED: electric drive, AC: air conditioning, HVB: high voltage battery)

The FC-PT is arranged in a hybrid configuration, where the fuel cell (FC) powers the 95-kW electric drive (ED) while a 14-kWh high voltage battery (HVB) is utilized as energy buffer. In this configuration the HVB can compensate the slow dynamic response of the FC by providing power to the ED when sudden changes of load occur. At the same time, the HVB allows driving and work at low loads as only power source, as the FC does not output power lower than 9 kW. The HVB can also be externally charged through an alternate current type 2 charging cable.

### 3.2 Thermal System Layout

One of the challenges in the FCTRAC is the thermal system's development. In contrast with internal combustion engines (ICEs) where waste heat is released at high temperature and partially through exhaust gases, in HVB and FC waste heat is produced at lower temperature and needs to be dissipated via coolant circuit, with an increasing demand on the cooling performance. The comparison in [26] between a FC-PT and a $H_2$ engine shows the power flux at the same rated power output: whereas the H$_2$ engine has ca. 40 % of waste heat removed via cooling water and ca. 60 % released through exhaust gases, the fuel cell powertrain's waste heat can be dissipated only through the cooling circuit. This increases the demand for heat exchange surface and new configurations to dissipate the waste heat, for example through the deployment of fans. Although fans account for a larger auxiliary power demand, they are necessary in tractors, as waste heat must be dissipated at heavy duty operation and low speed during field work.

The five cooling circuits of FCTRAC are depicted in Fig. 4. The oil cooling circuit (b) and the air conditioning circuit (d) are already present in the donor vehicle, whereas the FC cooling circuit (c), the high voltage (HV) components' cooling circuit (a) and the HVB cooling (e) are designed for the FC-PT.

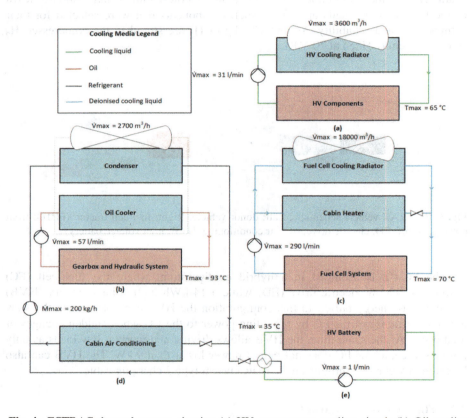

**Fig. 4.** FCTRAC thermal system circuits: (a) HV components cooling circuit (b) Oil cooling circuit (c) Fuel cell cooling circuit (d) Air conditioning circuit (e) HVB cooling circuit

The HVB cooling circuit is a liquid-cooled circuit, where the inlet coolant temperature must be kept under 35 °C for optimal operation. Therefore, the HVB circuit was designed in a nested configuration with the air conditioning circuit of the donor vehicle as depicted in Fig. 4, so that full operability is guaranteed even at ambient temperatures of 35 °C. With the higher refrigeration demand in the air conditioning circuit from the HVB waste heat, a fan is going to be installed on the air conditioning circuit condenser of the donor vehicle, which is going to be serially mounted with the oil cooler. An electrified air conditioning compressor is going to substitute the mechanically powered compressor of the donor vehicle.

The FC cooling system is a liquid-cooled circuit, where the coolant is a mixture of deionized water and FC ethylene glycol with conductivity lower than 5 μS/cm. Compared to standard ICEs coolant, low conductivity is a key requirement for the FC cooling system, where it is crucial to avoid current leakage and short circuit. Moreover, for optimal operation the FC temperature must be kept in a specific range. On one side, the FC coolant outlet temperature shall not exceed 82 °C to avoid water evaporation and consequently the dry-off of the membranes. On the other side, FC coolant inlet temperature shall not be lower than 30 °C to prevent water condensation that will block the gas diffusion layer and thus inhibit the transport of the reactants to the membrane. On the top of that, the temperature difference between inlet and outlet has to be kept minimal to generate a smooth temperature distribution over the stack. In terms of amount of waste heat, the FC cooling circuit has to dissipate the largest waste heat in the tractor and it was designed to compromise simplicity and cooling performance. As outlined by [27] in the comparison of various PEMFC systems, large scale PEMFCs (power output approximately higher than 10 kW) tend to be either liquid-cooled or use two-phase coolant. Although two-phase cooling has the advantage of less heat exchange surface, special condensers for deionized water raise the cost and the complexity of the circuit. Therefore, the liquid-cooled architecture was chosen as the most suitable architecture. As a matter of fact, other PEMFC applications on the market like Toyota Mirai [28] and prototypes like MAN PEMFC City Bus [29] deploy a similar liquid-cooled circuit with fans.

The HV components cooling circuit is a liquid-cooled circuit, where the power electronics components, including the ED, are cooled down. These components require different mass flows, temperature levels and pressure drops, hence the cooling circuit has been designed with a distributor supplying five main cooling lines where the components are mounted accordingly.

### 3.3 Thermal System Model

For the design of the thermal system, a thermal model of the tractor has been implemented. A longitudinal dynamic model of the donor vehicle in MATLAB®/Simulink® has been deployed for the determination of torque and engine speed demand to follow the target vehicle speed. This vehicle model was extended with the FC-PT that was parameterized by using among others the characteristic curves of FC and ED, efficiency maps of the power electronics components, and the capacity and the inner resistance of the HVB. The vehicle model also includes a transmission model, where the Power-Take-Off (PTO) of the tractor is considered. The waste heat of the vehicle components was calculated based on the efficiency of the respective operating point and used as input for a stationary model of the thermal system implemented in KULI. Thus, the vehicle model and the thermal model were coupled and co-simulations were carried out. The heat transfer at the radiators is modeled by:

$$\dot{Q} = \dot{m}_{IM} c_{p,IM} (T_{IN,IM} - T_{IN,OM}) \phi \qquad (1)$$

$\dot{Q}$ ... *radiator heat transfer,*
$\dot{m}_{IM}$ ... *mass flow of inner circuit medium,*
$c_{p,IM}$ ... *heat capacity of inner circuit medium,*
$T_{IN,IM}$ ... *inlet temperature of inner circuit medium,*
$T_{IN,OM}$ ... *inlet temperature of outer circuit medium,*
$\phi$ ... *operating characteristic*

The operating characteristic is a parameter that considers the effect of the turbulence on the heat exchange. For the radiators model, this parameter is defined experimentally as a function of the Reynolds number of the inner circuit medium and outer circuit medium.

The volume flow of the outer circuit medium is calculated from the measurement maps of the fan. In the modelling process of the heat exchange, conservative assumptions have been made: as the tractor speed during operation is generally low, the cooling effect of the airstream has been neglected.

As input for the simulation, vehicle speed and load profiles from recorded missions of the donor vehicle were used as representative operating points of the tractor. These points define target vehicle speed and variable load profiles on the three possible power ports of the tractor: power take-off, drawbar power, hydraulic power. As an example, one of the used profiles is depicted in Fig. 5, where the target vehicle speed is normalized to the maximum target speed and the hydraulic and drawbar power are normalized to the total power.

**Fig. 5.** An example of the used vehicle speed and load profiles

For the considered profile and ambient temperature 35 °C, the total maximum waste heat to be dissipated by thermal system is 209 kW, where the fuel cell accounts for the highest amount with 167 kW. The air conditioning circuit condenser dissipates 10 kW waste heat (where 2 kW from the HVB) and the HV cooling radiator 20 kW. The remaining 12 kW are taken out by the oil circuit.

The simulation results in Fig. 6 show the HV cooling circuit and FC cooling achieve to keep the coolant temperatures below the thresholds. The FC cooling system is the component that restricts the operation of the tractor, as the cooler inlet and outlet temperature almost reach the defined maxima. However, even by deploying a simple on/off control strategy of the FC radiator fans, the permissible FC coolant temperatures are not exceeded.

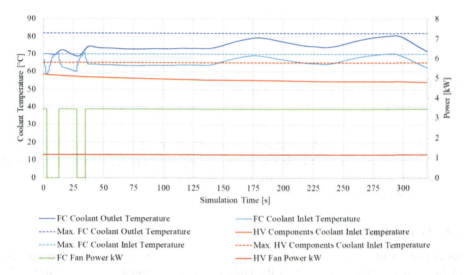

**Fig. 6.** Coolant temperatures in the considered speed and load profile

The effect of the ambient temperature on the FC coolant outlet temperature has been further investigated along with the power output. The steady-state simulation points in Fig. 7 show the full operability of the FC is guaranteed up to 35 °C ambient temperature even with worst-case conditions of stand-still and maximum power output through the PTO. In this configuration, the fans are constantly activated with maximum power 3.5 kW and the FC coolant outlet temperature can be kept under 82 °C. The ambient temperature was varied stepwise between 35 and 45 °C along with different PTO power output.

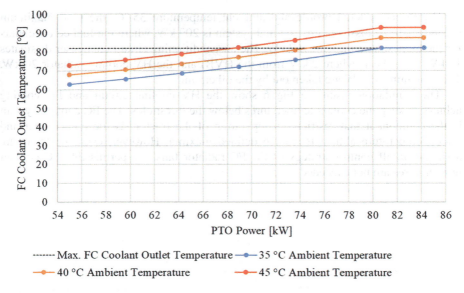

**Fig. 7.** FC coolant outlet temperature at stand-still with variation of ambient temperature and PTO power output

In the light of the made assumptions and the obtained results from the thermal model, it can be concluded the design of the cooling circuit concept meets the defined temperature requirements. Therefore, this concept has been selected and it will be further validated and integrated in the real tractor.

### 3.4 Vehicle Packaging

As already explained in Sect. 3.2, the increase of waste heat results in a partial rise of the heat exchange surface and the addition of new components, for example the fans compared to the diesel-powered donor vehicle. For this reason, the design of the vehicle packaging with the FC-PT has been a challenging task for the compact design of the tractor, which is a crucial design factor as operator's view must not be restricted. Including the additional components, the total volume increase of the packaging amounts ca. 3400 L compared to the donor vehicle. Therefore, due to the limited space in the engine bay, the CHSS and the FC cooler with the fans are going to be placed over the cabin on a support frame fixed to the chassis, as shown in Fig. 8. In this configuration ca. 2500-L space was obtained.

The ED and FC are going to be placed under the hood, whereas the HVB and HV components are going to be installed under the operator cabin. The HV components cooler is going to be situated in the left side of the vehicle. The right side of the tractor is going to be equipped with the air conditioning condenser and the oil cooler serially mounted with the fan.

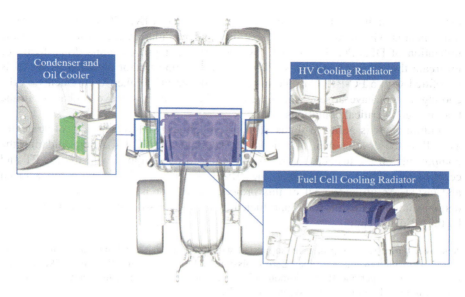

**Fig. 8.** Thermal system packaging schematic

### 3.5 Future Steps

First, efficient control strategies of the powertrain and thermal systems are going to be designed in model-based environment, then implemented and tested to guarantee an efficient and safe operation of the tractor. For this purpose, the ICE Control Unit will be removed and a dedicated Hybrid Control Unit will be developed and integrated in the existing vehicle control unit architecture. Along with this activity, the FC-PT components is in the process of being set up on the testbed, validated and successively integrated in the donor vehicle. In the last step, the vehicle is going to be certified as a prototype in view of demonstrations on public areas where it is going to be tested and fueled by the green $H_2$ of the "BioH$_2$Modul".

## 4 Conclusion and Outlook

In this paper, the solution "FCTRAC" for the zero-emission mobility in the agricultural sector was presented. For this goal, a broader vision has been adopted, comprising the development both of a FC tractor from an existing diesel vehicle and of an input-flexible production plant for high-purity $H_2$ from different feed gases. Aiming at a circular economy approach, a symbiotic interconnection between the vehicle and its operating environment was built.

Within the scope of this project, a plant for decentralized and biomass-derived $H_2$ production, the so-called "BioH$_2$Modul", has been developed. The necessary main process units for high-purity $H_2$ production from FIXB-PG have already been implemented with regard to input-flexibility and modularity, thus also enabling the utilization of DFB-PG, BG, or DG by adding further process units downstream or upstream the main process units. Parallelly, an existing tractor powered by diesel is retrofitted with a FC-PT. With the installation of the FC, higher requirements on the cooling system have arisen, which were fulfilled by an innovative packaging in the tractor even in critical conditions.

Future work is going to focus on the development of efficient control strategies as well as the development of a dedicated HCU for the FC-PT. After validating the components on the testbed, the tractor is going to be equipped with the FC-PT and certified as prototype. Meanwhile, the "BioH$_2$Modul" is going to be commissioned and tested. As ultimate goal of this project, the tractor is going to be deployed in demonstrations on public areas and fueled by the biomass-derived green $H_2$ produced by the "BioH$_2$Modul".

**Acknowledgments.** This project is supported with funds from the Climate and Energy Fund and implemented in line with the "Zero Emission Mobility" program. We would like to thank Christian Rathberger and Rolf Salomon of Magna Powertrain Engineering Center Steyr for the support in modeling the thermal system via KULI.

# References

1. World Energy Outlook. https://www.iea.org/reports/world-energy-outlook-2021 (2021). Accessed 9 Mar 2022
2. Technology Roadmap Delivering Sustainable Bioenergy. https://www.ieabioenergy.com/blog/publications/technology-roadmap-delivering-sustainable-bioenergy/. Accessed 9 Mar 2022
3. Adams, P., Bridgwater, T., Lea-Langton, A., Ross, A., Watson, I.: Greenhouse Gas Balances of Bioenergy Systems, 1st edn. Academic Press (2018). ISBN: 978-0-08-101036-5
4. Kaltschmitt, M., Hartmann, H., Hofbauer, H.: Energie aus Biomasse, 3rd edn. Springer, Berlin (2016). ISBN: 978-3-662-47437-2
5. Ryckebosch, E., Droullion, M., Vervaeren, H.: Techniques for transformation of biogas to biomethane. Biomass Bioenerg. **35**, 1633–1645 (2011). https://doi.org/10.1016/j.biombioe.2011.02.033
6. Ajhar, M., Travesset, M., Yuece, S., Melin, T.: Siloxane removal from landfill and digester gas – A technology overview. Biores. Technol. **101**, 2913–2923 (2010). https://doi.org/10.1016/j.biortech.2009.12.018
7. Simone, M., Barontini, F., Nicolella, C., Tognotti, L.: Gasification of pelletized biomass in a pilot scale downdraft gasifier. Beioresource Technology **116**, 403–412 (2012). https://doi.org/10.1016/j.biortech.2012.03.119
8. Müller, S., Fuchs, J., Johannes, S.C., Benedikt, F., Hofbauer, H.: Experimental development of sorption enhanced reforming by the use of an advanced gasification test plant. Int. J. Hydrogen Energy **42**, 29694–29707 (2017). https://doi.org/10.1016/j.ijhydene.2017.10.119
9. Abschlussbericht Monitoring Biogas II. https://www.dvgw.de/medien/dvgw/forschung/berichte/g1_03_10.pdf. Accessed 9 Mar 2022
10. Liu, K., Song, C., Subramani, V.: Hydrogen and Syngas Production and Purification Technologies. Wiley (2010). ISBN: 9780471719755

11. Bley Junior, C., Niklevicz, R. R., Alves, H. J., Frigo, E. P., Frigo, S. M., Coimbra-Araújo, C. H.: Overview of hydrogen production technologies from biogas and the applications in fuel cells. Int. J. Hydrogen Energy **38**, 5215–5225 (2013). https://doi.org/10.1016/j.ijhydene.2013.02.057
12. Du, Z., Liu, C., Zhai, J., Guo, X., Xiong, Y., Su, W., He, G.: A review of hydrogen purification technologies for fuel cell vehicles. Catalysts **11** (2021). https://doi.org/10.3390/catal11030393
13. Aasadnia, M., Mehrpooya, M., Ghorbani, B.: A novel integrated structure for hydrogen purification using the cryogenic method. J. Clean. Prod. **278** (2021)
14. Bennson, J., Celin, A.: Recovering Hydrogen — and Profits — from Hydrogen-Rich Offgas. Chem. Eng. Prog. **114**, 55–59 (2018). https://doi.org/10.1016/j.jclepro.2020.123872
15. Biogas upgrading technologies – developments and innovations. https://www.ieabioenergy.com/wp-content/uploads/2009/10/upgrading_rz_low_final.pdf. Accessed 9 Mar 2022
16. Woolcock, P.J., Brown, R.C.: A review of cleaning technologies for biomass-derived syngas. Biomass Bioenerg. **52**, 54–58 (2013). https://doi.org/10.1016/j.biombioe.2013.02.036
17. Milne, T.A., Evans, R., Abatzoglou, N.: Biomass gasifier "tars": their nature, formation, and conversion. Technical Report, National Renewable Energy Laboratory. Colorado (1998). https://doi.org/10.2172/3726
18. Tonpakdee, P., Hongrapipat, J., Siriwongrungson, V., Rauch, R., Pang, S., Thaveesri, J., Messner, M., Kuba, M., Hofbauer, H.: Influence of solvent temperature and type on naphthalene solubility for tar removal in a dual fluidized bed biomass gasification process. Curr. Appl. Sci. Technol. **21**, 751–760 (2021). https://doi.org/DOI.14456/cast.2021.60
19. Loipersböck, J., Weber, G., Rauch, R., Hofbauer, H.: Developing an adsorption-based gas cleaning system for a dual fluidized bed gasification process. Biomass Convers. Biorefin. **11**, 85–94 (2021). https://doi.org/10.1007/s13399-020-00999-1
20. Paethanom, A., Nakahara, S., Kobayashi, M., Prawisudha, P., Prawisudha, K.: Performance of tar removal by absorption and adsorption for biomass gasification. Fuel Process. Technol. **104**, 144–154 (2012). https://doi.org/10.1016/j.fuproc.2012.05.006
21. Weinlaender, C., Neubauer, R., Hauth, M., Hochenauer, C.: Adsorptive hydrogen chloride and combined hydrogen chloride–hydrogen sulphide removal from biogas for solid oxide fuel cell application. Adsorpt. Sci. Technol. **36** (2018). https://doi.org/10.1177/0263617418772660
22. Petersson, A.: Biogas cleaning in The Biogas Handbook. Woodhead Publishing Limited (2013). ISBN: 9780857094988
23. Tansel, B., Surita, S.C.: Managing siloxanes in biogas-to-energy facilities: Economic comparison of pre- vs post-combustion practices. Waste Manage. **96**, 121–127 (2019). https://doi.org/10.1016/j.wasman.2019.07.019
24. Tran, V.T.L., Gélin, P., Ferronato, C., Mascunan, P., Rac, V., Chovelon, J.-M., Postole, G.: Siloxane adsorption on activated carbons: Role of the surface chemistry on adsorption properties in humid atmosphere and regenerability issues. Chem. Eng. J. **371**, 821–832 (2019). https://doi.org/10.1016/j.cej.2019.04.087
25. Thomas, C.E.: Fuel cell and battery electric vehicles compared. Int. J. Hydrogen Energy **34**, 6005–6020 (2009). https://doi.org/10.1016/j.ijhydene.2009.06.003
26. Kufferath, A., et al.: $H_2ICE$ powertrains for future on-road mobility. In: 42. Internationales Wiener Motorensymposium, Vienna (2021)
27. Eun Jung Choi: Jin Young Park, Min Soo Kim: Two-phase cooling using HFE-7100 for polymer electrolyte membrane fuel cell application. Appl. Therm. Eng. **148**, 868–877 (2019). https://doi.org/10.1016/j.applthermaleng.2018.11.103

28. Nonobe, Y.: Development of the Fuel Cell Mirai. IEEJ Trans. Electr. Electron. Eng. **12**, 5–9 (2017). https://doi.org/10.1002/tee.22328
29. MAN PEMFC City Bus. https://www.researchgate.net/publication/242755697_HYDROGEN-FUELED_BUSES_THE_BAVARIAN_FUEL_CELL_BUS_PROJECT. Accessed 8 Feb 2022

# Systems Engineering for Fuel Cell Vehicles: From Simulation to Prototype

Daniel Ritzberger($\boxtimes$), Alexander Schenk, and Falko Berg

AVL List Gmbh, Graz, Austria
{daniel.ritzberger,
alexander.schenk,falko.berg}@avl.com

**Abstract.** In order to develop complex systems that meet the customer expectations, the systems engineering methodology has been successfully applied in many industries. With the goal to break down the customer targets into tangible requirements on sub-system and component level, holistic simulation methodologies are heavily encouraged, as thereby the complex interactions of systems and components can be investigated. In this work, the system engineering methodology and holistic simulation concept is showcased for a fuel cell heavy-duty vehicle, and highlights the workflow and the required toolset from concept to functioning prototype.

**Keywords:** Fuel cell system · Heavy duty · Systems engineering · Model-based design

## 1 Introduction

The global climate crisis and the need for action towards a sustainable future, including strategic energy independence, poses a significant challenge for the mobility and transportation sector. In order to reduce emissions, the electrification of the automotive industry is inevitable. Battery electric vehicles (BEV) have already proven their technological readiness and are continuously increasing their market share for personal transportation and light duty applications. However, for the electrification of heavy duty vehicles, pure battery electric power train configurations are currently limited to urban traffic and last-mile logistics distribution [1] due to their weight and limited battery autonomous range [2]. As an alternative, fuel cell systems are regarded as a viable option for the electrification of heavy-duty vehicles. In a fuel cell system, hydrogen is used as the main energy carrier, which is directly converted to electric energy by the use of fuel cells. As thereby the energy storage and the electrochemical power generation is spatially separated (unlike for batteries), the driving range can be scaled independently from the fuel cell system rated power (much like with a conventional ICE configuration). Additionally, refueling times are also significantly lower than charging times for pure BEV drivetrains, which helps in reducing the

---

© Der/die Autor(en), exklusiv lizenziert an Springer Fachmedien Wiesbaden GmbH, ein Teil von Springer Nature 2023
A. Heintzel (Hrsg.): ATZLive 2022, Proceedings, S. 105–118, 2023.
https://doi.org/10.1007/978-3-658-41435-1_9

106      D. Ritzberger et al.

total cost of ownership (TCO), which is regarded as a main driver for the heavy-duty market [3].

Understanding the top-level requirements and developing technical solutions, which are optimal with regard to their intended application, is the main focus of systems engineering. Thereby, the top level requirements are continuously decomposed into sub-system and component requirements, while still considering their interconnection and functional integration.

In order to develop and verify a system design and additionally, to manage the complexity between lateral and vertical integration of systems and components and their functional interdependency, the development workflow is increasingly rooted in model-based development methodologies and simulation [4].

In this work the model-based development process for a fuel cell system in a heavy-duty application is presented. To facilitate the systems engineering approach, and in order to front-load design decisions, a holistic simulation toolchain has been developed. The resulting fuel cell system design, and its integration into a heavy-duty truck will be shown and an outlook on future topics will be given.

## 2    Model-Based Design: From Vehicle Targets to Component Requirements

For the model-based system development, a holistic simulation framework is desired. Holistic is thereby to be understood in two ways:

1. The simulation framework should facilitate different levels of abstraction and modeling complexities of each component and sub-system
2. The simulation framework should allow a lateral and vertical integration of components and sub-systems to investigate their interdependencies

Regarding the first point: the modeling environment should be flexible with respect to the model complexity, as the level of abstraction should be a reflection of the current task at hand. Or conversely: There is no "one-size-fits-all simulation". A simulation model is constantly evolving throughout a development project. Thereby modeling assumptions – which are necessary in the first place to resolve obvious development conundrums – are continuously replaced by in-depth modeling if needed. The other way around, e.g. starting with the most complex and integrated model in order to answer all the potential use-cases and eventualities that might come up during development, is on the one hand unreasonable, and on the other hand, inefficient. Adhering to the principle of parsimony, a model should always be as simple as possible and as complex as necessary. Regarding the second point: functional requirements cannot be formulated by investigating a sub-system or a component in an isolated environment. It is the interaction and integration of a multitude of systems that in turn result in a successful product design. As such, it is necessary to investigate and track how design decisions influence each other in order to convergence to an optimal design.

For the fuel cell system design in a heavy-duty application, the top level of the simulation framework (Fig. 1) consists of the vehicle simulation.

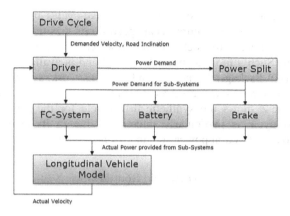

**Fig. 1.** Holistic simulation framework: Vehicle Level

The corresponding top-level requirement are usually formulated with respect to specific drive-cycles or mission profiles, that are to be fulfilled by the vehicle and usually comprise of a demanded vehicle velocities (under specific road and ambient conditions). The discrepancy between actual vehicle velocity and demanded velocity is used in the driver model, which, in a technical sense, is a suitable control algorithm, whose output is in the most general case the demanded propulsion power by the vehicle. The overall propulsion power demand is an input to the power split strategy, which in turn distributes the overall power demand between the relevant sub-systems (fuel cell system, battery and brakes). The actual power, provided by the individual sub-systems, is then forwarded into the vehicle model in order to calculate the actual velocity, thus, completing the simulation loop.

In the remainder of this work a few selected topics will be discussed in detail to solely focus on the methodology – all other involved building blocks are typically treated equally and would require a much more in-depth discussion.

## 2.1 Drive Cycles: Considering Real Driving Conditions

The demanded drive-cycle of the vehicle constitutes a crucial input into the simulation framework and should reflect the top-level requirements of the vehicle. For heavy-duty applications, several standardized drive-cycles are readily available (e.g., WLTP, Vecto [5]). However, by only using a handful of standardized drive-cycles as the main input into the simulation framework, there is a high risk of over-engineering the system requirements with respect to these drive-cycles, which most likely will result in a non-optimal system design when considering real-world operation. As such, a multitude of drive-cycles under (ideally) real driving conditions are desired, in order to assess design decisions in a statistical meaningful way. But since experimentally generating real-world driving data under all relevant operating conditions is, on the one hand, extremely cost intensive and on the other hand unreasonable, especially in an early concept phase of a development project, suitable alternatives are in high demand.

One possible solution is to synthesize the drive-cycles using geographic route data (e.g. elevation profile, curvature, legal speed limits, position of traffic lights, stop and yield signs) together with stochastic variations emulating the variance of on-road traffic conditions. Such a tool has been developed and is available in the AVL Smart Mobile Solutions™ Package. In the remainder of this work, synthesized drive-cycles based on real-world routes are used as the main input into the holistic simulation framework. Figure 2 shows the synthetic drive-cycle for the route from Vienna to Graz. This drive-cycle is particularly interesting due to its challenging elevation profile leading to high traction power demand for heavy-duty vehicles.

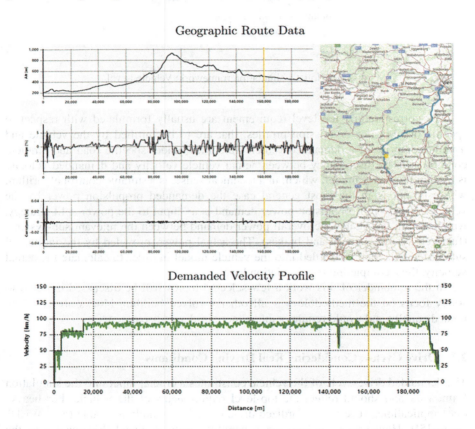

**Fig. 2** Drive Cycle synthesis from Vienna to Graz using AVL Smart Mobile Solutions™. The upper three plots show the geographic route data and the bottom plot the resulting demanded velocity profile for the heavy-duty truck

### 2.2 Vehicle Simulation

With the macroscopic vehicle parameters given in Table 1, the overall demanded mechanical traction power is determined through simulation.

**Table 1.** Vehicle simulation macroscopic parameters

| Vehicle Mass | 40,000 kg |
|---|---|
| Drag Coefficient | 0.6 |
| Reference Area | 9.6 m$^2$ |
| Roll Resistance | 0.0065 |

In Fig. 3, the histogram of the mechanical traction power is shown and the proposed power levels of the relevant sub-system of the power-train are given as a visual indication.

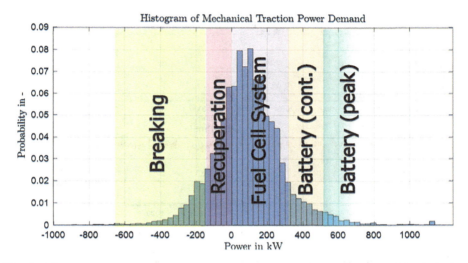

**Fig. 3.** Histogram of the demanded mechanical traction power and proposed load sharing between the relevant sub-systems of the heavy-duty vehicle

With a fuel cell system net power of 300 kW, 90% of the demanded mechanical traction power can be supplied. By considering an auxiliary battery with a capacity of 70 kWh and a C-Rate of 2, the demanded traction power coverage can be increased to 96.5%. However, providing traction power above the maximum power rating of the fuel cell system leads to a decrease in the State of Charge (SoC) of the Battery. As such, operating beyond the maximum power rating of the fuel cell system cannot be maintained indefinitely.

By analyzing a multitude of synthetic drive-cycles for the target application (in the target region with corresponding legal ramifications) a statistically significant basis is formed that allows for an informed decision on the sub-system sizing, i.e.: defining the trade-off between the coverage of the fuel cell system power vs. peak-power vs. battery size, and therefore: capital cost and package size of the fuel cell system.

With the requirements of the relevant sub-systems refined, the demanded fuel cell system net power can be determined via the vehicle simulation. Note, that therefore a power-split strategy needs to be defined. This strategy however is not unique and can be used to shift the cycle load between sub-systems, influencing the overall power train efficiency and degradation of the fuel cell stack and battery [6]. In the following, a naïve SoC-Hold strategy is applied, i.e. the battery state of charge is maintained at a pre-defined level. With such a strategy, the batteries main task is to smooth the transient power demands otherwise imposed on the fuel cell system and to provide peak power demands.

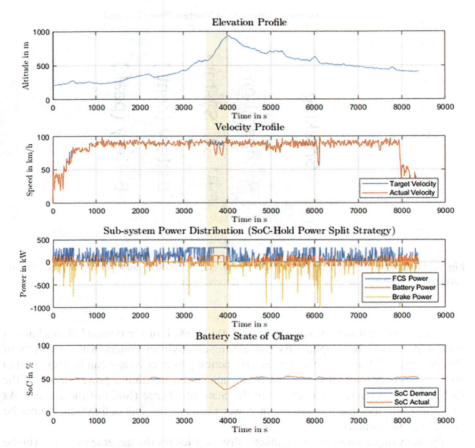

**Fig. 4.** Top level vehicle simulation and resulting power demand to the relevant sub-systems (FCS, Battery and Brake). The highlighted area corresponds to a peak power scenario (i.e. demanded mechanical traction exceeds maximum power of the fuel cell system) due to the significant road inclination

In Fig. 4. the elevation profile, the demanded and actual vehicle velocity, the power distribution to the relevant sub-systems as well as the battery state of charge is shown. By design, the average power demand is provided by the fuel cell system and as such, the Battery SoC can be maintained for the majority of the drive-cycle. However at around 3500s, a challenging elevation climb is required from the heavy-duty vehicle, resulting in peak power operation above the maximum power of the fuel cell system. As such the State of Charge is depleting. As previously mentioned, analyzing such outlier scenarios (and assess their probability) can enable an informed decision in terms of balancing the sub-system requirements. However, it is to be noted, that an "optimal" powertrain configuration is not determined by purely taking into account the technical considerations, i.e. the degree to which a demanded drive-cycle power can be fulfilled. Naturally, this would lead to an oversized powertrain in order to achieve a maximum coverage of demanded traction power, even for the most unlikely outlier scenario. A more economically reasonable development target would be to balance the Total Cost of Ownership (TCO) with respect to the robustness of outlier driving scenarios.

## 2.3 Fuel Cell System Simulation

With the relevant requirements for the powertrain sub-systems defined and a chosen power split strategy to balance the load between fuel cell system and battery, the resulting net power demand from the fuel cell system for the drive-cycle is determined, with its histogram shown in Fig. 5.

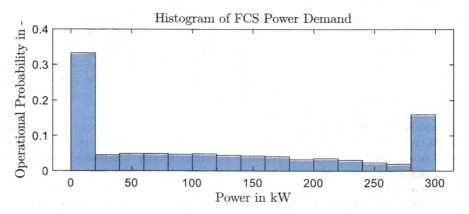

**Fig. 5.** Histogram of the demanded net fuel cell system power for the drive-cycle used

A significant operational probability is visible for the full load operation at 300 kW, which is indicative for the heavy-duty applications. Subsequently, a high efficiency at maximum fuel cell power is desired. On the one-hand, this motivates the use of a turbine in the fuel cell system exhaust path, as for full load operation a significant amount of enthalpy is available in the cathode exhaust stream which could be harnessed via the isentropic expansion of a turbine. On the other-hand, the resulting net efficiency of the fuel cell system can be influenced by the sizing of the fuel cell stack.

By increasing the cell-count, the peak power of the fuel cell system is increased and, due to the efficiency characteristic of a fuel cell system, the efficiency at the design point (i.e. 300 kW in this case) is thereby increased. A visual representation thereof can be seen in Fig. 6.

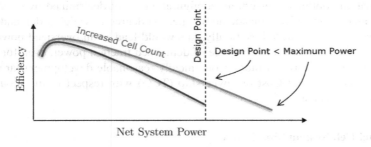

**Fig. 6.** Representation of the interactions between net system power, net efficiency and cell count of the Fuel Cell Stack(s). Increasing the cell count leads to higher net efficiencies in a design power point (e.g. 300 kW)

This highlights another trade-off for the fuel cell system design. By allocating the maximum amount of cells in the available package space of the fuel cell stack the highest net efficiency would be achievable. However, also the capital cost of the fuel cell system is thereby increased. Again, for the selection of the stack cell count, the TCO should be considered. As for heavy-duty applications, fuel costs dominate the total cost of ownership, and therefore an "over-sizing" of the fuel cell stack with respect to its design point is more feasible then for example a passenger car, where the initial cost constitutes the main market driver.

Another possibility to increase the overall efficiency of the fuel cell system, would be the use of an advanced, predictive power split strategy [7]. Essentially, road predictions of the desired route (elevation profile and traffic conditions) can be used to predict the changes in potential energy of the vehicle which can be utilized to adapt the demanded SoC of the battery to maximize to overall powertrain efficiency (e.g. SoC is sufficiently high at the beginning of a climb at a minimum at the apex) by the means of optimal control theory. Due to these predictive SoC control, peak loads can be redistributed to lower (more efficient) load demands.

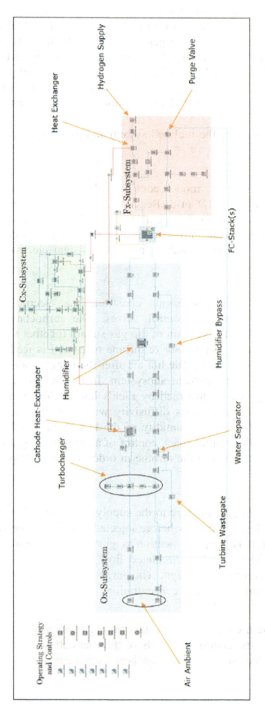

**Fig. 7.** Fuel Cell System Simulation in AVL CRUISE$^{TM}$ M

In the next step, the detailed fuel cell system simulation is developed, for which AVL CRUISE™ M is utilized. An overview of the included sub-systems and components can be seen in Fig. 7. In general, AVL CRUISE™ M is a multi-physics system simulation tool considering first principles (e.g. the established laws of physics) on a low dimensional (0D/1D) spatial grid. The resulting simulation model is dynamic and real-time capable, and therefore also utilized in a Model in the Loop (MiL) and Hardware in the Loop (HiL) environment for detailed parameterization and control concept studies.

The main subsystems of the fuel cell system and its components are summarized in the following.

**Fuel Cell Stack**

The transient fuel cell stack model considers the mass and species balance of cathode and anode along a 1D grid. Reaction of Species on the catalyst layer as well as species transport across the membrane (i.e. diffusion of water and nitrogen, and electro-osmotic) are taken into account as well. In order to couple the transient thermodynamic states to current and voltage, a consistent electrochemical model is used [8].

**Air Subsystem**

The air subsystems main components constitute the electric turbocharger, the humidifier, heat-exchanger and control valves. The turbocharger considers the isentropic compression and expansion of the gas stream together with the component specific operational maps (compressor & turbine map) and is feedback controlled to supply the desired air mass flow. The hot air after the compressor is cooled using an intercooler connected to the coolant subsystem. In the humidifier, the water transport from wet to dry side is governed by Fick's law of diffusion along a 1D grid. The dependency of the membranes diffusivity with respect to its water content is thereby considered. The relative humidity of the humid air supplied to the cathode inlet is actively controlled via feedback control of a humidifier bypass valve. A water-separator is implemented before the turbine in order to prevent any water droplets to impact the turbine.

**Hydrogen Subsystem**

After reducing the hydrogen pressure to the supply-line level (12bar) the hydrogen is pre-warmed using a heat-exchanger before injected into the anode recirculation loop via an ejector/injector. Hydrogen injection is governed by controlling the pressure at the anode inlet. The overall recirculated flow in the anode loop is determined passively by the ejector/injector design. The purge valve is feed-forward controlled in order to achieve a desired hydrogen concentration.

**Cooling Subsystem**

The cooling subsystem consists of a coolant pump, three-way control valve, a main coolant loop through the cooling channels of the fuel cell stack and parallel coolant loops connected to the air- and hydrogen heat exchanger. Either an ideal heat-sink or an air/liquid heat exchanger with an actively controlled radiator fan is used to dissipate the thermal heat load of the fuel cell system.

**Operating Strategy and Controls**

The operation of the fuel cell system is either ensured by using a simplified control logic in the simulation model (i.e. cascade of PID controllers and simplified current dependent operating strategy maps) or by coupling the simulation model to the actual control software in MATLAB/Simulink. Real-time coupling between AVL CRUISE™ M and the Simulink based control software is achieved via AVL Model. CONNECT™.

## 2.4 Holistic Simulation: Investigating Interactions

With the fuel cell system simulation model available, a direct coupling to the longitudinal vehicle simulation model is possible in order to investigate the interdependencies across multiple horizontal and vertical integration levels. Thereby, the holistic simulation toolchain – from drive-cycle to the component level – is effectively established. This holistic simulation toolchain allows for an in-depth evaluation of components and sub-systems and supports the interactive derivation of system and component requirements as well as its virtual validation.

As an example, in Fig. 8 the operation conditions of the compressor during the simulated drive-cycle are projected onto the compressor map.

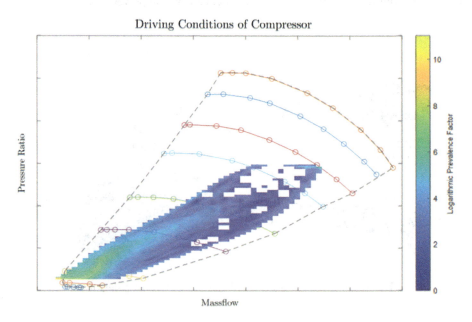

**Fig. 8.** 2D Histogram of the operating conditions of the compressor during the drive-cycle, projected onto the compressor map

These operational conditions of the compressor are the result of the highly interconnected and interdependent simulation toolchain, e.g. from the drive-cycle to the vehicle model and power-split strategy, to the operating strategy of the fuel cell system and lastly to the control strategy and actuation of the Balance of Plants components such as the compressor. Thereby, the regions of high operating probability of the turbocharger are clearly visible. Additionally, the width of the operation area (and therefore the proximity to surge and choke line) is directly influenced by matching of the mass flow and pressure control requirements. With the simulation of a fuel cell vehicle under real-world scenarios, the suitability of components and operating strategies can be front-loaded onto the virtual development phase of a project, gaining invaluable insights early on in order to make informed decisions on engineering questions. And since any decision regarding a function, sub-system or component can have a significant impact to its adjacent counter-part (on a horizontal as well as vertical level) a holistic simulation is required in order to develop an optimal fuel cell system with respect to the customer requirements.

## 3 Resulting Fuel Cell System

The derivation and validation of requirements in a virtual environment is not a one-way street but in close cooperation with the feedback regarding the availability of package, space, components and costs. This again underlines the desired flexibility of any simulation solution in order to efficiently investigate and react with respect to new information during the development of the fuel cell system power train.

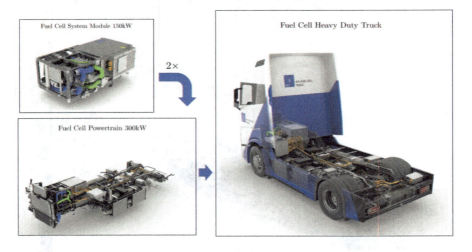

**Fig. 9.** Integration of the fuel cell system into the heavy duty truck

The resulting mechanical integration and packaging of the fuel cell system can be seen in Fig. 9. With currently available automotive components and an AVL in house developed fuel cell stack, the integration strategy has been to develop a modular fuel cell system approach: One Fuel Cell System module provides 150 kW net power and in order to fulfill the desired power demand of the heavy-duty truck, two fuel cell systems are packaged into the power train on order to provide the desired 300 kW net power.

Due to the highly integrated development workflow and the significant degree of virtual engineering in a holistic simulation environment, the resulting fuel cell system exhibits a superior power density as with respect to competitor systems available on the market (Fig. 10)

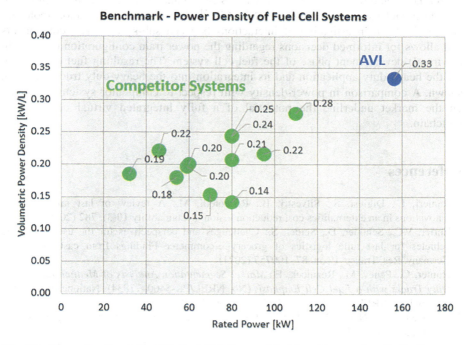

**Fig. 10.** Power density of the AVL Fuel Cell System Module with respect to fuel cell systems available on the market

## 4 Conclusion & Outlook

In this work, a holistic simulation toolchain – from the drive-cycle to the component level – has been demonstrated for the development of heavy-duty fuel cell systems. This holistic toolchain is necessary in order to develop, keep track and virtually validate the system requirements in order to meet the customer demands. A high level of flexibility is thereby required in order to minimize development time and effort. The synthesis of drive-cycles from geographic route data in AVL Smart Mobile Solutions™ has been shown in order to emulate real-driving, which is prerequisite in order to develop an optimal power train configuration for a specific application. The significant interactions between vehicle simulation, power split strategy and fuel cell system simulation have been highlighted. A detailed fuel cell system simulation model has been developed using AVL CRUISE™ M and coupled with the longitudinal vehicle model using AVL Model.CONNECT™. The resulting simulation toolchain enables in-depth investigation of interactions between sub-systems and components and allows for informed decisions regarding the power train configuration early on in a (virtual) development phase of the fuel cell system. The resulting fuel cell system for the heavy-duty application and its integration into the heavy-duty truck has been shown. A comparison in power-density with respect to competitor systems available on the market underline the potential of a fully integrated virtual development toolchain.

## References

1. Ranieri, L., Digiesi, S., Silvestri, B., Roccotelli, M.: A review of last mile logistics innovations in an externalities cost reduction vision. Sustainability **10**(3), 782 (2018)
2. Ehrler, V.C., Schöder, D., Seidel, S.: Challenges and perspectives for the use of electric vehicles for last mile logistics of grocery e-commerce–Findings from case studies in Germany. Res. Transp. Econ. **87**, 100757 (2021)
3. Hunter, C., Penev, M., Reznicek, E.: *Market Segmentation Analysis of Medium and Heavy Duty Trucks with a Fuel Cell Emphasis* (No. NREL/PR-5400–77834). National Renewable Energy Lab (NREL), Golden, CO (United States) (2020)
4. Haberfellner, R., Nagel, P., Becker, M., Büchel, A., von Massow, H.: Systems Engineering, p. 5. Springer, Cham (2019)
5. Franco, V., Delgado, O., Muncrief, R.: *Heavy-Duty Vehicle Fuel-Efficiency Simulation: A Comparison of US and EU Tools. ICCT white paper* (2015)
6. Ferrara, A., Zendegan, S., Koegeler, H.M., Gopi, S., Huber, M., Pell, J., Hametner, C.: Optimal calibration of an adaptive and predictive energy management strategy for fuel cell electric trucks. Energies **15**(7), 2394 (2022)
7. Pell, J., Schörghuber, C., Pretsch, S., Schreier, H.: Optimized Operating Strategies for Fuel Cell Trucks. ATZ Heavy Duty Worldwide **15**(1), 26–31 (2022)
8. Kravos, A., Ritzberger, D., Tavčar, G., Hametner, C., Jakubek, S., Katrašnik, T.: Thermodynamically consistent reduced dimensionality electrochemical model for proton exchange membrane fuel cell performance modelling and control. J. Power Sources **454**, 227930 (2020)

# Sustainability Assessment of an Integrated Value Chain for the Production of eFuels

Jana Späthe$^{(\boxtimes)}$, Manuel Andresh, and Andreas Patyk

Institute for Technology Assessment and Systems (ITAS),
Karlsruhe Institute of Technology (KIT), Karlsruhe, Germany
`{jana-spaethe,manuel.andresh,andreas.patyk}@kit.edu`

**Abstract.** EFuels are discussed as an alternative to fossil fuels for several applications within the transport sector. In this paper, a sustainability assessment is presented considering an integrated process chain, which is developed and investigated in the project Kopernikus P2X. The assessment consists of a Techno-Economic Assessment (TEA), an environmental Life Cycle Assessment (LCA) and a literature review about the future production potential. Thereby, the LCA and TEA are based on the results presented in the 3rd roadmap of the Kopernikus P2X project. In addition, the focus is on the implications of the electricity input. The results show that there is an environmental benefit in case of global warming, if renewable electricity is used. Additionally, the improvement can be magnified by increasing the efficiency of the power plants through higher capacity factors at the individual sites. Inexpensive renewable electricity and high full load hours are important for economic considerations. Apart from that, it is likely that policy measures will be required to make production and use of eFuels economical compared to fossil equivalents. The review of potential analyses showed that the majority of the production of eFuels for German demand will most likely not be in Germany. In addition, it can be assumed that the quantity of eFuels that can be produced will not be sufficient for all branches of the transport sector. This necessitates prioritizing those modes of transportation where direct electrification is not possible.

**Keywords:** Sustainability assessment · Heat integration · Power-to-X fuel

## 1 Introduction

### 1.1 Efuels in the Context of Transport Transition

In June 2021 the German government adopted new binding climate targets. To achieve these, greenhouse gas neutrality must be reached by 2045 [1]. Therefore, emissions must fall in all sectors, which is already happening well in some of them for example in the electricity generation industry. Nevertheless, the emissions of the transport sector still increase [2].

© Der/die Autor(en), exklusiv lizenziert an Springer Fachmedien Wiesbaden GmbH, ein Teil von Springer Nature 2023
A. Heintzel (Hrsg.): ATZLive 2022, Proceedings, S. 119–129, 2023.
https://doi.org/10.1007/978-3-658-41435-1_10

The use of eFuels, also named Power-to-X (PtX) fuels, which are produced with electricity out of $CO_2$ and water is discussed in this context. In principle, these fuels can replace the fossil fuels used today, gasoline, diesel and kerosene, without major changes in vehicles and the required infrastructure. Therefore, eFuels are discussed for several applications: from air transport, ship transport and trucks to passenger cars. The technical suitability and the environmental benefit for the different applications is nevertheless not enough for decision-making. Economic and social aspects have to be considered as well. As eFuels need electricity and fresh water to be produced, their availability is an equally important factor.

This paper focuses specifically on a heat integrated process chain, developed and investigated in the Kopernikus project P2X. The process chain consists of a Direct Air Capture (DAC) plant, a high temperature (HT) co-electrolysis, a Fischer-Tropsch (FT) synthesis and the equipment for the final processing. The process chain and its characteristics are described in detail in the following subchapter. The technical development in the Kopernikus P2X project is accompanied by a roadmapping process including life cycle assessments (LCA) and techno-economic assessments (TEA). These assessments, which have been published in the roadmap 3.0 [3] are the basis for the analyses considered in the following.

This previous research has already shown that the impact of the electricity source on the results of environmental and economic assessment of the eFuel production value chain is high. Therefore, the aim of this paper is to show, which environmental and economic minimum criteria have to be fulfilled by the electricity input in order to produce an eFuel, which emits less greenhouse gases while being economic sensible. In addition, a non-exhaustive literature review about the production potential of eFuels is part of this work. According to these assessments, conclusions can be drawn, which parts of the transport sector can reasonably be covered by eFuels in the future.

Another challenge in using renewable electricity is that most energy sources, such as solar and wind, are only available on a fluctuating basis. Therefore, it is important to investigate how different numbers of full load hours affect greenhouse gas emissions respectively the costs of production.

## 1.2  Heat Integrated Process Chain

Currently, innovative production ways for synthetic fuels are developed, one of them at the KIT as part of the Kopernikus project P2X. The special features of this FT based fuel production concept are that all of the production steps, from the DAC plant to the HT co-electrolysis to the FT-synthesis and the final processing, are highly integrated and the applied technologies, e.g. FT-synthesis with microstructured apparatus, are suitable for small-scale container plants. The integration of these plants leads to efficiency gains, e.g. due to heat recovery. Especially the heat released from the exogenous process step of FT synthesis is utilized by the DAC plant. In addition, residual gases, a mixture of unreacted synthesis gas, is separated and recycled to the HT co-electrolysis, which results in a high C utilization rate of the process chain. Separated wastewater could also be reused in the future after water treatment. The potential of this optimization approach is not yet considered. Since waste heat and the carbon-containing gases can be efficiently reused in the process chain, an efficiency of almost just under 50 % LHV (lower heating value) 55 % HHV (higher heating value) can be achieved. [3]

## 2 Methodology

In this paper an environmental as well as an economic assessment of the production and combustion of E-fuels is done based on the LCA and the TEA of the 3rd roadmap of the Kopernikus P2X project [3]. A special focus is thereby on the influence of the electricity input. Only the scenarios for 2050 are considered, which are further specified in the roadmap. An industrial scale production has been assessed with a 100 MW HT co-electrolysis as starting point for the simulation. The data has been provided by various technical partners. The methodology of LCA and TEA, as well as the approach used in this study, are described in more detail in the following subsections. These main analyses are complemented by a subsequent chapter that provides own calculations and a non-exhaustive literature review on the availability of renewable electricity for eFuels production.

### 2.1 Life Cycle Assessment

LCA is a method for evaluating the environmental impacts of products over their whole life cycle. It is standardized and defined in ISO 14040 and 14044 [4, 5]. In order to get an idea, which kind of electricity input is suitable for the production of eFuels in the first place, the focus of this paper is on global warming. Therfore, the impact category global warming potential of the impact assessment method ReCiPe 2016 (H) is used. The assessment is done with the ecoinvent database version 3.6 in the system model "cut-off by classification".

In the roadmap 3.0 it has been shown that for the analyzed case and under the in the roadmap further described assumptions the construction of the plants and auxiliaries are responsible for only about one tenth of the emissions, when the production is running continuously (8000 h per year) [3]. The majority of the impacts come from the electricity and heat input. For this study, therefore, the influence of the construction of the plants is taken from the roadmap and only the $CO_2$ emission intensity of the electricity mix is varied. This range of emissions per kWh of electricity input is then compared with scenarios for the German grid mix [3] as well as the electricity production technologies wind and PV. The first reference is based on calculations for PV and wind electricity production in Germany presented in the roadmap [3]. This range for emissions from PV or wind electricity production shows the influence of the construction of the plants. For example, in the future, cement and steel production will continue to develop and reduce their emissions. In the same time, the emissions generated by the production of PV or wind power will then also decrease. Additionally, the greenhouse gas impact per kWh PV or wind electricity is highly dependent on the solar irradiation respectively the wind volume at the plant location. Therefore, this paper also includes calculations for Morocco, which has a high solar irradiation and Iceland, which has a high wind volume [6]. The calculations are based on a tool that accounts for capacity factors and thus solar irradiance or wind volume at a given location. The underlying data for this also comes from the ecoinvent database. The capacity factors are taken from [7], who provide hourly capacity factors for PV and wind based on [8] and [9]. The general capacity factor was adjusted to the location. In this case, it was assumed that, despite the increased output, no additional maintenance or other input is required at the power plant. The tool has been described in [10]. Apart from these factors, the greenhouse gas impact of renewable electricity production is also dependent on recycling of the power plant materials [11]. The range given in the diagrams below is based on several publications concerning life cycle emissions of wind electricity [3, 11–13].

Additionally, as described above, the influence of full load hours is investigated. Therefore, a scenario with 4000 h per year is calculated, which is roughly equivalent to the production of offshore wind power in the North/Baltic Sea and hybrid use of PV and wind power in North Africa [14]. At this point, a simplified assumption was made that the plant follows the fluctuations of the renewable power source, regardless of technical feasibility. As the lifetime of the production facilities is assumed to be constant (20 years), the influence of the construction per product is doubled. The influence of a possibly necessary transport is not examined, since this is dependent on the location of the plant and thus cannot be represented in the investigation type selected in this study.

As reference the production and combustion of fossil diesel is considered. However, there is only a small difference between the environmental impact of production and combustion of the various fossil fuels. The results are modelled using the ecoinvent 3.6 database, whereby the electricity and heat inputs are varied in the same way as for the eFuel.

## 2.2 Techno-Economic Assessment

The techno-economic assessment is a method to assess the production costs of a product. The net production costs (NPC), are the sum of investment costs (CAPEX) and operating costs (OPEX). In the assessment presented in the roadmap 3.0 [3] the data basis is a process simulation and information on investment costs provided by the technical partners. This is supplemented by surcharge factors based on the methodology described in [15]. Similar to the environmental assessment, in this study, the results for construction and auxiliaries are taken from the roadmap while the cost of electricity is being varied. Heat inputs have a minor share compared to electricity and are assumed to also be provided by electricity here.

Even though all results are presented over a certain range of electricity costs, possible scenarios for the German electricity mix from the P2X roadmap as well as electricity costs of exemplary renewable energies are included in the figures for orientation. In this way, other possible power inputs can also be quickly compared if their costs are known.

In addition, the influence of a reduced number of operating hours is also investigated. For this purpose, 4000 h are considered in addition to 8000 h. Therefore, the CAPEX and the indirect OPEX were doubled, since half of the output is produced per year, but the lifetime of the plants was assumed to be constant. Costs for transport of the fuel to Germany is not included. As reference the average of product procurement costs for diesel in 2021 and February 2022 is presented [16].

## 3 Results

### 3.1 Global Warming Impact of the Production of eFuels

As described above, the focus of this paper is on providing a detailed analysis on the impact of the electricity input. In Fig. 1 the results for the net greenhouse gas (GHG) emissions for two full load hour scenarios (8000 and 4000 h) of the eFuel as well as the result for the reference fossil diesel are presented. The net GHG emissions for the eFuel are calculated by subtracting the $CO_2$ emissions released during the combustion, because the same amount of $CO_2$ is taken from the atmosphere by the DAC plant at the beginning of the production process chain. Like this, only the $CO_2$ emissions of the production are considered for the FT-fuel. In contrast to this, the reference is the production and combustion of fossil diesel.

As the influence of the electricity input is around 90% of the results, the difference between the two full load hour scenarios is small. For both scenarios, the break-even point is around 150 g $CO_2$-equivalent per kWh, where the life cycle GHG emissions of the use of the two fuel options are the same.

For orientation different possible electricity inputs have been implemented. Two scenarios for 2030 and 2050 have been developed during the Kopernikus P2X project and presented in the roadmap 3.0. The scenarios "DE 2030" and "DE 2050" incorporate the currently no longer valid German government's climate protection targets of 2019 for emissions reductions in 2030, 2040, and 2050. According to this, greenhouse gas emissions are to be reduced by at least 80 to 95 % until 2050 compared to 1990. The presented scenario "DE 2050" does not include fossil electricity sources anymore. [3]

When an electricity mix, which includes even more fossil electricity sources than "DE 2030" is used, using the FT-fuel has a larger impact on global warming than the use of the fossil fuel. This is the case for the electricity mix of today, which has a global warming impact of around 485 g $CO_2$-equivalent per kWh [3] as well as for every fossil electricity source itself.

The picture is much better when using a renewable source of electricity such as wind or solar power. In particular, the use of wind power with a high efficiency results a savings potential of around 85% (4000 h) to 90% (8000 h) for PtX fuel compared to the fossil equivalent. The impacts of PV electricity are higher and both, wind and solar power, are dependent on the production location and the associated capacity factors. This shows that the efficiency of the renewable electricity source is an important factor for the magnitude of environmental benefit of using eFuels instead of fossil fuels.

124     J. Späthe et al.

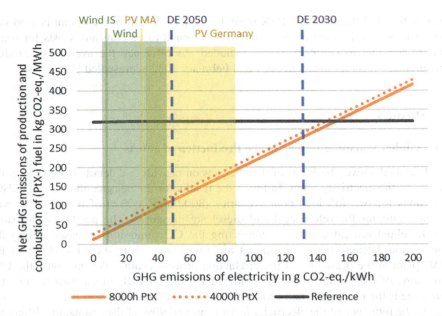

**Fig. 1.** Net greenhouse gas emissions of synthetic FT-fuel compared to fossil fuel. Two scenarios with different full load hours and are included. Additionally, two electricity scenarios as well as two renewable electricity sources are plotted for orientation

### 3.2 Economic Aspects of the Production of eFuels

In order to find out, if eFuels are a sustainable alternative for fossil fuels, it is also important to consider the economic aspect. In Fig. 2 the production costs of PtX-fuel are presented together with the global market price for kerosene. Both are plotted in the diagram on a range of power generation costs.

Regardless of the cost of electricity, there is no intersection between the two alternatives. The production costs of FT-fuel in the presented 2050 scenario of the roadmap 3.0 are thus always higher than the current kerosene price. Nevertheless, the difference, which has to be overcome, is increasing with increasing electricity price. For comparison three electricity production cost scenarios for German grid electricity in 2020, 2030 and 2050, which are taken from the roadmap 3.0, are shown. As described above, the greenhouse gas emissions of the current electricity mix are too high, so the focus is here on the scenarios for 2030 and 2050. These electricity production costs are within the cost range, which can be found for German wind electricity in 2050 in the literature [14]. They lead to around 4.6 times larger production costs compared to the costs of fossil diesel in 2021.

Cheaper electricity prices could be found for example for PV electricity produced in North Africa, Middle East and in general in countries with a high global radiation in 2050 [14]. This reduces the difference to about 2.8 times in the 8000 h scenario compared to the procurement costs of fossil diesel 2021.

The influence of reducing the full load hours in the 4000 h scenario is higher in the economic calculations than in the environmental assessment. The reason for this is the higher share in production costs of CAPEX and indirect OPEX compared to the environmental influence of facility building. A lower number of full load hours significantly increases the cost difference and makes economical operation less likely.

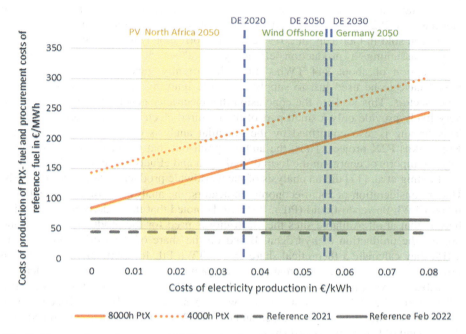

**Fig. 2.** Presented are the costs of PtX- and reference fuel production plotted on a range of electricity production costs. For orientation three cost scenarios for German grid electricity as well as costs for PV electricity in North Africa and offshore wind electricity in Germany in 2050 are marked

### 3.3 Production Potential and Availability of Renewable Electricity

For the prevention of climate change it is important to find solutions which prevent greenhouse gas emissions while being cost-effective. In addition, it is important to make sure that upscaling is possible and also social aspects are considered. There are already several publications in the literature that deal with the production potential of eFuels [3, 17, 18]. In this chapter, some of these considerations are summarized and supplemented by own thoughts.

For the production of eFuels two inputs are important: renewable energy and freshwater. In Germany, the limiting factor is the availability of renewable electricity [3]. During the ongoing discussions about the electricity source for eFuels the use of so called "surplus" electricity is often mentioned. This unused electricity exists, when renewable power plants provide more electricity than is needed at that moment in that area. The problem of using it for PtX production is at one hand the availability of the current in a certain area and the resulting low possible full load hours per location and on the other hand the low total amount of "surplus" electricity. Additionally, there are more options for using this electricity through e.g. power-to-heat or storage systems. [14]

A total amount of 250 TWh of renewable electricity have been generated in Germany in 2020 [19]. Compared to that, around 10.2 million tonnes of kerosene have been sold in Germany in 2019 [20]. This corresponds to 122 TWh of kerosene, which, assuming a production efficiency of about 50%, results in electricity consumption of about 244 TWh. Thus, all currently generated renewable power would have to be used just to supply air traffic ignoring the increasing demand in other sectors. This leads to the conclusion that electricity generation in Germany can very likely provide only a small part of the required electricity for eFuel production. This finding is also confirmed by the potential analysis carried out as part of the Kopernikus P2X project. Thus, the authors point out that Germany will continue to rely on imports of energy in the form of hydrogen and eFuels in the future [3].

Pfennig et al. [17] did an analysis of the worldwide production potential of eFuels. Under consideration of socio-economic indicators, the analysis showed a potential of 69.000 TWh Power-to-liquid (PtL). It should be noted that the underlying assumption here is that at all available sites the electricity is used for the production of PtL and not for the production of hydrogen. Based on the share of the world population in 2018, they calculate a theoretical share of 643 TWh PtL for Germany. [17] This can now be compared with the German demand again. For the aviation kerosene demand described above, the comparison now looks much better. In automobile traffic, 198 TWh of diesel and 226 TWh of gasoline were consumed in 2019 [21]. Together with kerosene demand, this results in a demand of 546 TWh for eFuels without considering the demand of trucks, buses and ships.

If the mentioned worldwide production capacity is reached is highly depending on the possible expansion dynamic for renewable energies in the possible production countries. Additionally, the potential reduction of greenhouse gas emissions through the shutdown of coal-fired power plants in the countries itself is higher than the use of the electricity for the production of PtX. [17]

In summary, the considerations presented show that renewable electricity – in Germany as well as worldwide – will not be available in unlimited quantities in the future and should therefore be used as efficiently as possible.

# 4 Discussion

In order to find out, whether eFuels are a suitable alternative for a specific means of transport, various aspects must be addressed, which have been examined in this paper.

First of all, lower greenhouse gas emissions over the whole life cycle are essential to be considered as an alternative to fossil fuels. Section 3.1 showed that this is not always the case. The use of renewable electricity is substantial, and here, too, there are differences in life-cycle emissions. For example, solar power has higher greenhouse gas emissions per kWh than wind power. Additionally, it is important for efficiency reasons that the plant location has a high irradiation respectively sufficient wind volume. Nevertheless, eFuels with both renewable electricity options have a clear advantage over fossil fuel and the potential analyses show that, in terms of quantity, every renewable source of electricity is necessary to cover at least part of the transport sector. Other renewable electricity sources that have not been explored in detail here are also worth considering. A review of further impact categories in the context of a detailed LCA is also important, since the production of eFuels also requires fresh water, which is not available everywhere without limits. In addition, the expansion of renewable power plants requires the use of materials and resources.

The economic analysis also showed that electricity costs have a significant impact on the total cost of eFuels. Reasonable renewable electricity prices are thus a basic prerequisite for using eFuels on a larger scale. In areas with high solar irradiation or good wind conditions, renewable electricity can also be produced more cost-effectively through higher efficiency. Especially due to their good transportability, eFuels are suitable to bring this energy to Germany [17]. Nevertheless, it is likely that in 2050 the remaining costs, excluding electricity costs, are still already higher than the current price of fossil fuels. In addition, costs are sensitive to low full-load hours, which increases the gap with the price of fossil equivalents. Low electricity prices and high full load hours are thus essential for economic operation of the plants. This could also be favored in the future by rising prices of crude oil as well as political measures.

In addition, it is important to mention that measures required to compensate for the fluctuating temporal availability of wind or solar power were not considered. Flexible operation of the plant, which results in less full load hours and/or additional electricity storage for increasing the full load hours might be possible solutions here.

The potential analyses presented in Sect. 3.3 show that renewable electricity will probably not be available in sufficient quantities in the future to cover all areas of the transport sector with eFuels. It will therefore be necessary here to prioritize those areas in which direct electrification e.g. through the use of batteries, which is associated with higher efficiency, is possible.

# 5 Conclusion

Various aspects of a utilization of eFuels in the transport sector have been addressed in this paper. In terms of emissions of climate-damaging gases, eFuels have an advantage over fossil fuels if the $CO_2$ is extracted from the air beforehand and renewable electricity is used as input. On the other hand, it is likely that policy measures will be required to make production and use economical compared to fossil equivalents. In any case, attractive renewable electricity prices are important in order to make economical production feasible.

The review of potential analyses showed that the majority of the production of eFuels will most likely not be in Germany. In addition, it can be assumed that the quantity of eFuels that can be produced will not be sufficient for all branches of the transport sector. It will therefore be necessary to prioritize those means of transport for which direct electrification is not an option. Last but not least, the availability of eFuels will mainly depend on the expansion of renewable electricity production.

For the purposes of the roadmap and this paper, eFuels have been examined on the basis of a specific value chain. If a different technology is investigated, the results and interpretation may differ. It should also be noted that the study is subject to uncertainties. Further research is needed, especially on how to deal with the fluctuation of renewable electricity sources during production.

# References

1. Bundesministerium für Wirtschaft und Klimaschutz: Deutsche Klimaschutzpolitik. https://www.bmwi.de/Redaktion/DE/Artikel/Industrie/klimaschutz-deutsche-klimaschutzpolitik.html (2022). Accessed 28 Mar 2022
2. European Environment Agency (EAA): Greenhouse gas emissions from transport in Europe. https://www.eea.europa.eu/data-and-maps/indicators/transport-emissions-of-greenhouse-gases-7/assessment (2021). Accessed 22 Oct 2021
3. Kopernikus Projekt P2X: Optionen für ein nachhaltiges Energiesystem mit Power-to-X-Technologien. https://www.kopernikus-projekte.de/projekte/p2x/#roadmaps (2021). Accessed 22 Oct. 2021
4. DIN EN ISO 14040:2009-11, Umweltmanagement_- Ökobilanz_- Grundsätze und Rahmenbedingungen (ISO_14040:2006); Deutsche und Englische Fassung EN_ISO_14040:2006. Beuth Verlag GmbH, Berlin (2006)
5. DIN EN ISO 14044:2018-05, Umweltmanagement_- Ökobilanz_- Anforderungen und Anleitungen (ISO_14044:2006_+ Amd_1:2017); Deutsche Fassung EN_ISO_14044:2006_+ A1:2018. Beuth Verlag GmbH, Berlin (2018)
6. International Aspects of a Power-To-X Roadmap. A report prepared for the World Energy Council Germany (2018)
7. Renewables.ninja. https://www.renewables.ninja/. Accessed 28 Mar 2022
8. Pfenninger, S., Staffell, I.: Long-term patterns of European PV output using 30 years of validated hourly reanalysis and satellite data. Energy (2016). https://doi.org/10.1016/j.energy.2016.08.060
9. Staffell, I., Pfenninger, S.: Using bias-corrected reanalysis to simulate current and future wind power output. Energy (2016). https://doi.org/10.1016/j.energy.2016.08.068
10. Andresh, M.: Sustainability Requirements Analysis of Power-to-X-Fuels in the Aviation Sector, 18 November 2021
11. Tremeac, B., Meunier, F.: Life cycle analysis of 4.5 MW and 250 W wind turbines. Renew. Sustain. Energy Rev. (2009). https://doi.org/10.1016/j.rser.2009.01.001
12. Bonou, A., Laurent, A., Olsen, S.I.: Life cycle assessment of onshore and offshore wind energy-from theory to application. Appl. Energy (2016). https://doi.org/10.1016/j.apenergy.2016.07.058
13. Wiedmann, T.O., Suh, S., Feng, K., Lenzen, M., Acquaye, A., Scott, K., Barrett, J.R.: Application of hybrid life cycle approaches to emerging energy technologies–the case of wind power in the UK. Environ. Sci. Technol. (2011). https://doi.org/10.1021/es2007287
14. Perner, J., Unteutsch, M., Lövenich, A.: Die zukünftigen Kosten strombasierter synthetischer Brennstoffe. Agora Verkehrswende, Agora Energiewende & Frontier Economics (2018)

15. Peters, M.S., Timmerhaus, K.D., West, R.E.: Plant design and economics for chemical engineers, 5th edn. McGraw-Hill chemical engineering series. McGraw-Hill, Boston (2004)
16. en2x: Preiszusammensetzung des Verbraucherpreises für Dieselkraftstoff. https://en2x.de/service/statistiken/preiszusammensetzung/ (2022). Accessed 28 Mar. 2022
17. Pfennig, M., Bonin, M. von, Gerhardt, N.: PtX-Atlas: Weltweite Potenziale für die Erzeugung von grünem Wasserstoff und klimaneutralen synthetischen Kraft-und Brennstoffen. Teilbericht im Rahmen des Projektes: DeV-KopSys (2021)
18. Perner, J., Bothe, D.: Internationale Aspekte einer Power-to-X Roadmap (2018)
19. Umweltbundesamt: Deutlich weniger erneuerbarer Strom im Jahr 2021. https://www.umweltbundesamt.de/presse/pressemitteilungen/deutlich-weniger-erneuerbarer-strom-im-jahr-2021 (2022)
20. Mineralölwirtschaftsverband (MWV): Jahresbericht 2020. https://en2x.de/wp-content/uploads/2021/11/MWV_Mineraloelwirtschaftsverband-e.V.-Jahresbericht-2020.pdf (2020)
21. Deutsches Zentrum für Luft- und Raumfahrt e. V. (DLR); Deutsches Institut für Wirtschaftsforschung (DIW); Kraftfahrt-Bundesamt: Verkehr in Zahlen 2021/2022. https://www.bmvi.de/SharedDocs/DE/Publikationen/G/verkehr-in-zahlen-2021-2022-pdf.pdf?__blob=publicationFile (2021)

# Creating and Sustaining User Engagement in Bidirectional Charging

Franziska Kellerer[1]([✉]), Johanna Zimmermann[2],
Sebastian Hirsch[1], and Stefan Mang[1]

[1] Institute CENTOURIS, University of Passau, Passau, Germany
{franziska.kellerer,sebastian.hirsch,
stefan.mang}@uni-passau.de
[2] Chair of Marketing and Innovation,
University of Passau, Passau,
Germany
johanna.zimmermann@uni-passau.de

**Abstract.** Encouraging users to buy and continuously engage in bidirectional charging are crucial for the long-term success of the technology. However, user research in the context of smart charging has mainly focused on investigating overall perceptions and factors motivating consumers to buy the technology. In this article, we aim to take a more specific perspective on user acceptance of the technology by investigating both their preferences with regard to the design of the business model of bidirectional charging as well as the design of app feedback mechanisms for creating long-term user engagement. Our findings from study 1 reveal that financial aspects (i.e., preferred contractor, type of compensation, willingness to initially invest in the technology) constitute an important factor impacting user participation in bidirectional charging. In the long run, however, users' non-financial motivation must also be addressed. Therefore, study 2 sheds light on how to foster users' charging behavior by implementing gamified app feedback (i.e., financial, social, and efficient energy use). With this article, we contribute to user research which has been largely neglected in the highly technology-focused field of bidirectional charging.

**Keywords:** Bidirectional charging · Customer motivation · User engagement

## 1 Introduction

Bidirectional charging enables Plug-In electric vehicles (PEVs) to not only draw electrical power from the grid or photovoltaic system for their high-voltage battery when plugged into a compatible charging station or wall box, but also to reverse the process and feed it back into the grid or household [1]. As such, the intelligent integration of PEVs into grid operations has great potential to counteract imbalances

© Der/die Autor(en), exklusiv lizenziert an Springer Fachmedien Wiesbaden GmbH, ein Teil von Springer Nature 2023
A. Heintzel (Hrsg.): ATZLive 2022, Proceedings, S. 130–139, 2023.
https://doi.org/10.1007/978-3-658-41435-1_11

between supply and demand of energy which, for example, helps reducing $CO_2$ and the dependence on fossil energy sources by making better use of renewable energy sources. The successful implementation of technological innovations such as bidirectional charging is closely linked to and largely dependent on user acceptance and participation [2]. So far, prior user research in the context of intelligent charging systems has mainly focused on investigating factors that encourage users to participate in bidirectional charging (e.g., reasons to invest in the technology): Besides sociodemographic aspects and personal characteristics of the customer, *financial compensation, non-financial aspects* (e.g., more efficient energy consumption and environmental protection) as well as *social incentives* (e.g., hedonic aspects, social comparison) have been found to be driving users' acceptance of bidirectional charging [3]. Prior studies have found that financial compensation is the most important driver of user participation in bidirectional charging [4].

However, while user research has mainly focused on overall perceptions and motivational factors to use bidirectional charging, it so far missed to shed light on the question, how factors that motivate users to invest in bidirectional charging technologies (e.g., financial compensation) should be integrated in the design of its underlying business model in order to meet user preferences. Further, to the best of our knowledge, there is no research investigating the crucial question of how to keep customers engaged in the bidirectional charging technology in the long run (i.e., motivating them to plug in their cars and make charging settings). This holistic perspective is important as only if used continuously, the advantages of the bidirectional charging technology fully apply for all parties involved.

As part of the Research Project *"Bidirektionales Lademanagement – BDL"* ("Bidirectional Charging Management – BCM") funded by the German Federal Ministry for Economic Affairs and Climate Action, we aim to fill this void and address the following research questions in this article.

- How can underlying business models of the bidirectional charging technology be designed to increase user acceptance of bidirectional charging?
- How can customers be motivated through financial, social and environmental feedback to continuously use bidirectional charging after purchase?

To pursue our research goals, we conducted two studies building on results from prior studies conducted as part of the BCM project. Within study 1, we aim to provide a better understanding of specific user preferences depending on prior PEV-experience regarding financial compensation in the V2G (Vehicle-to-Grid) context. The study focused on the design of business models for bidirectional charging by comparing different configurations of business model parameters (i.e., preferred contractor, type of compensation, willingness to invest). Adopting a gamification perspective and following a multi-method approach, study 2 then investigates how app feedback (i.e., financial, social and environmental) can be designed in a motivating manner to encourage consumers to continuously use bidirectional charging.

In the subsequent sections of the remaining article, we provide background information for each study before explaining its design, procedure and results. We then discuss the main findings of our studies and conclude by highlighting avenues for further research.

## 2 Study 1: Design of Potential Business Models for Bidirectional Charging

In order to be successful in the long run, bidirectional charging needs to provide benefits for users after its implementation to compensate potential obstacles of the technology (e.g., fear of battery degradation, uncertainty with regard to flexibility needs, increased costs) [5]. Former studies have shown that financial compensation represents the most important factor to motivate users' participation to invest in the technology [4], however, research so far mainly focused on the economic feasibility and potential rewards payoffs from a systemic perspective. We assume that financial compensation can only effectively foster user engagement if the business model meets their preferences.

This is in line with prior research arguing that, from a user perspective, one crucial barrier for the adoption of the bidirectional charging technology are poor business models [5]. Therefore, study 1 focuses on developing a better understanding of user preferences regarding important aspects of potential business models for bidirectional charging. Thus, we investigate what preferences users have regarding the contractor (i.e., one-stop-shop vs. freedom of choice regarding the provider for different components), how they want to be compensated (i.e., saving in their energy bill vs. separated payment) and what amount of money they would be willing to spend as an initial investment to acquire the technology.

### 2.1 Design, Participants, Procedure

In order to address the first research question and understand how V2G business models should be designed aiming at a high user acceptance, we conducted an online survey from January until February 2021. In total, we recruited 1196 participants representative of the German population with regard to age and gender ($M_{age} = 47.05$, 43.30 % female). The sample included a non-representative high share of plug-in electric vehicle (PEV) users ($n_{PEV\ users} = 264$), which allowed target group specific data analysis with regard to PEV-ownership (vs. non-ownership).

To introduce our participants to the technology, we presented a video outlining the concept of bidirectional charging technology in the context of the V2G use case. In this introduction the participants learned that the use of the technology has various advantages, including the potential to generate profits. We further assured, that individual mobility needs will not be threatened when using bidirectional charging by telling them that a "security"-state-of-charge will not be undercut by the system. Based on the assumption that high interest in the technology is crucial for the overall business model success, we first asked participants about their interest in different use cases of the bidirectional charging technology by integrating a closed-ended question which allowed for multiple responses (i.e., interest in Vehicle-to-Grid, interest in Vehicle-to-Home, I do not know, I am not interested).

Participants then were asked to state their preferences with regard to potential business model components. In particular, our participants indicated their preferences regarding the contractor for overall system (i.e., one-stop-shop for the acquisition of

Creating and Sustaining User Engagement ...        133

all system components vs. possibility to acquire all system components at separate dealers) as well as their preferred mode of compensation (i.e., savings in electricity tariff vs. separate monthly payout) using close-ended questions, which allowed for a single response. Next, we asked them to state their willingness to initially invest in the technology and the expected revenues from bidirectional charging. We integrated open text-fields where respondents could first state the amount of money they would spend as a one-time investment in the technology and, second, indicate what they would expect to earn from using bidirectional charging. Both questions allowed for no response. Additionally, we integrated a question regarding individual car ownership considering combustion engine as well as electric and alternative engine (e.g., hydrogen) types, which allowed us to analyze the participants answers with regard to their level of PEV-experience.

## 2.2 Results

*General interest in business model.* With regard to our question on participants' general interest in the technology, our results show an overall high willingness to engage in bidirectional charging. With 59.1 %, the majority of survey respondents stated, that they are generally interested to use the technology. We found that participants who already own a PEV indicated a higher interest to use bidirectional charging (79.9 %) in comparison to respondents, that do not own a PEV (53.2 %). Also, participants of both groups stated, that they consider the presented V2G use case scenario as generally interesting. Again, those respondents who own a PEV state a higher interest (47.8 %) than non-PEV users (26.6 %). These group differences in the willingness to engage in the bidirectional charging technology in general as well as in the V2G use case in particular proofed to be significant. We verified this finding by conducting a Qui-Square-Test (General interest: $\chi^2 = 62.48$, $p < .001$; Interest in V2G: $\chi^2 = 42.69$; $p < .001$).

*Business model design.* The evaluation of the respondents' preferences regarding the contractor for the overall system (i.e., consisting of the regenerative electric car, a bidirectional wallbox and the corresponding electricity tariff) revealed that users prefer a free choice of the certain supplier of the components over a one-stop-shop where they would get all required system components from one single provider. This result especially holds true for the group of users already owning a PEV (74.6 %) in comparison to the group of non-PEV owners (60.4 %,). We found these group differences to be significant ($\chi^2 = 17.94$; $p < .001$). Furthermore, we investigated how the revenues generated by bidirectional charging should be returned to the users. When being asked about their preferred form of financial compensation, more than half of the survey participants stated to prefer savings in their electricity expenses over a separate payoff, with no significant group differences to be found (63.3 % of PEV users; 58.0 % of non-PEV users; $p > .05$).

With regard to participants' willingness to initially invest into the bidirectional charging technology as well as their minimum expected annual savings, our results show that on average PEV owners are accepting significantly higher one-time investment expenses than non-PEV owners ($M_{PEV\ users} = 1{,}398.16$ €; $M_{non\text{-}PEV\ users} = 901.40$ €;

134    F. Kellerer et al.

$p < .001$). The different willingness to invest between the two groups is further illustrated in relation to the expected annual savings. Across all participants the stated expected annual savings were on average 311.96 € with no significant group differences ($M_{PEV\ users}$ = 323.82 €; $M_{non\text{-}PEV\ users}$ = 305.74 €; $p > .05$). Overall, the juxtaposition of maximum accepted one-time investment costs and expected annual savings indicate, that people who already own a PEV show a significantly higher willingness to invest.

*Additional results.* A repeated question on the importance of different motivational factors to use bidirectional charging was included, that asked to distribute an overall number of 100 points to several factors, dependent on their relative importance to the participant. In line with prior project results [4], financial compensation was again found to be of highest importance for both groups ($M_{PEV\ users}$ = 48.02; $M_{non\text{-}PEV\ users}$ = 55.77; $p < .001$). Interestingly, the study revealed, that environment and grid-related benefits are also important in that context, *especially for users who already own a PEV*. In the light of this finding we consider both financial and environmental motivational factors, as well as social factors in our second study, where we focus the question on how gamified feedback can create continuous user engagement in the bidirectional charging technology.

## 3   Study 2: Employing Gamified Feedback to Sustain Customer Engagement

An essential prerequisite for bidirectional charging management to allow for sustainable energy behavior is the continuous usage of the technology. This implies that consumers need to put time and effort into regularly plugging in their car and making their app settings to have their car ready when needed. So far, both research and practice have mainly focused on motivating consumers to buy the technology while the critical issue of encouraging users to continually use the technology has been largely neglected.

Prior research in fields of, for example, health, education and crowdsourcing [6] provides a promising approach to keep consumers engaged in a task, that is, *gamification*. In general, gamification *"refers to a process of enhancing a service with affordances for gameful experiences in order to support users' overall value creation"* [7, p. 25]. Motivational affordances can take different forms, e.g., avatars, challenges, badges or rewards [8]; they are increasingly implemented in different types of information systems (e.g., apps) to make information and feedback more appealing for users, while, at the same time, addressing their intrinsic or extrinsic motivations [6]. In the context of energy apps, research has already investigated, for example, the effectiveness of employing gamification to raise awareness of consumers' energy consumption habits [9] or to motivate energy-saving behavior [10].

Based on former research as well as prior studies conducted as part of the BCM project, we employ gamification to examine how feedback in an app for bidirectional charging should be designed in order to address users' motivations and engage them in long-term technology usage. This study served as a base for implementing gamified feedback in the BiLi-App which is developed within the BCM-project.

## 3.1 Design, Participants, Procedure

To pursue our research objective, we employed a multi-method approach consisting of three steps: *creativity workshop, expert evaluation* and *quantitative study*. First, we carried out a creativity workshop ($N = 8$) to generate ideas on how to design gamified feedback addressing the main motivations of bidirectional charging (i.e., financial, social and efficient energy use) identified in prior research [3]. Within this workshop, 24 drafts were prioritized by the participants (7 targeting financial, 10 targeting environmental and 7 targeting social motivations).

After transferring the sketches from the creativity workshop into the app design, six experts from the field of user research were asked to revise, reduce and improve the ideas from the creativity workshop. Despite giving feedback with regard to liking (1 "I like it very much" to 7 "I don't like it at all") and gamification degree (1 "Could be much more playful" to 7 "Could be much less playful"; 4 corresponds to "Is just right playful"), the experts gave suggestions for improvement within a free text field based on ISO 14915 heuristics as well as ISO 9241 for website/app development [11, 12]. Within the expert evaluation, the ideas generated within the creativity workshop were classified according to specific underlying concepts and the motivational affordances to illustrate the concepts were adapted. Two concepts per motivation were prioritized by the experts (see Table 1).

Finally, we recruited 240 participants ($M_{age} = 41.07$, 48.75% female) from a professional panel provider to evaluate the gamified feedback concepts. After being introduced to the bidirectional charging technology as well as a hypothetical incentive system developed by the researchers[1], each participant was randomly assigned to one concept of each motivational factor and evaluated them based on the short version of the *Attrakdiff* scale using a 7-point semantic differential scale to rate pragmatic quality (pq) and hedonic (hq) quality as well as attractiveness (att) of the concepts [13].

## 3.2 Results

With the help of the multi-stage process described in Sect. 3.1, different ideas for gamified feedback design (financial, social, efficient energy use) could be generated and aggregated into six relevant concepts, which are intended to stimulate user motivation for the permanent use of bidirectional charging. It should be noted that all concepts presented tended to score well (min. 4.38 in the overall evaluation),

---

[1] At the time of the study, there was no final incentive concept developed within the BCM project to design a reward system. Therefore, we implemented a fictional incentive concept using points as a motivational affordance that can be collected and exchanged for certain benefits (e.g., financial compensation).

136     F. Kellerer et al.

**Table 1.** Gamified Feedback Concepts in BCM (for a review of motivational affordances see [7])

| Financial Feedback Concepts and Gamified Design | |
| --- | --- |
| *Financial 1 (f1):* <br> *Underlying concept:* Collecting points to reach the next level and receive achievement badges which represent financial benefits <br> *Motivational affordances:* Points; levels; achievement badges | *Financial 2 (f2):* <br> *Underlying concept:* Setting an individual goal to reach a certain amount of points and receive financial benefits <br> *Motivational affordances:* Points; clear goals; performance stats |
| **Social Feedback Concepts and Gamified Design** | |
| *Social 1 (s1):* <br> *Underlying concept:* Setting point goals as a group to create a community spirit; illustration of individual contribution within the group <br> *Motivational affordances:* Points; cooperation; clear goals | *Social 2 (s2):* <br> *Underlying concept:* Collecting points to win a virtual race against other users (highest amount of points wins) to foster competition among users <br> *Motivational affordances*: Points; competition; performance stats |
| **Efficient Energy Use Feedback Concepts and Gamified Design** | |
| *Efficient Energy Use 1 (e1):* <br> *Underlying concept:* Displaying the share of time [in %] bidirectional charging mode is used relative to the total charging time/ week to illustrate efficient energy use (e.g., in terms of supporting grid stability, environmental protection or optimizing own energy consumption); color change, if the percentage share of bidirectional charging drops below 50 % of total charging time/week <br> *Motivational affordances:* Status bar; performance stats; warnings | *Efficient Energy Use 2 (e2):* <br> *Underlying concepts:* Displaying the amount of energy charged in bidirectional charging mode relative to the total amount of energy charged to illustrate efficient energy use (e.g., in terms of in terms of supporting grid stability, environmental protection or optimizing own energy consumption); color change, if the percentage share of bidirectional charging drops below 50% of total charged energy/day <br> *Motivational affordances:* Status bar; performance stats; warnings |

indicating that the methodology used in this research is well suited to develop concepts for gamified feedback design.

Overall, the financial feedback design performed best: In line with previous studies and our findings from study 1, participants are primarily motivated by financial incentives. The first financial feedback concept picked up on the level idea which is known from certain video games: By reaching a certain number of points, users could reach the next level which (in our example) was accompanied by a reduction of the electricity price per kWh. The second financial feedback concept ($M_{pqf2} = 5.04$; $M_{hqf2} = 4.85$; $M_{attf2} = 5.00$; $M_{overallf2} = 4.96$) was evaluated slightly better than the first ($M_{pqf1} = 5.12$; $M_{hqf1} = 4.80$; $M_{attf1} = 4.92$; $M_{overallf1} = 4.95$). It focused on the idea of allowing users to set individual goals (i.e., collecting points) and, thus, strive for personal progress and goal achievement.

Interestingly, the concept focusing on "competition with oneself" (*financial 2*; individual user goals and striving for personal progress) scored best among all concepts, while the social feedback design concept focusing on a competition with others scored the lowest ($M_{pgs2} = 4.39$; $M_{hqs2} = 4.37$; $M_{atts2} = 4.39$; $M_{overalls2} = 4.38$). The first social feedback concept focused on the idea of community achievements to create a sense of connectedness when pursing a common goal. The individual share in the achievement of a community goal also tends to have a motivating effect here (social comparison). The concept received the third best rating among all concepts ($M_{pgs1} = 4.75$; $M_{hqs1} = 4.73$; $M_{atts1} = 4.78$; $M_{overalls1} = 4.75$).

Finally, environmental feedback also plays an important role in motivating users' long-term engagement with the technology. The high similarity of the feedback concepts targeting efficient energy use is reflected in an almost equal scoring ($M_{pge1} = 4.80$; $M_{hqe1} = 4.52$; $M_{atte1} = 4.70$; $M_{overalle1} = 4.67$; $M_{pge2} = 4.77$; $M_{hqe2} = 4.53$; $M_{atte2} = 4.75$; $M_{overalle2} = 4.68$): While the first concept focused solely on illustrating the amount of bidirectional charging time relative to the overall charging time (in %; overall time also included time in immediate charging mode), concept 2 also included the amount of energy charged in bidirectional charging mode.


# 4   Summary and Discussion

In summary, both studies show that financial compensation is a core factor not only to motivate customers to invest into the technology, but also to keep them engaged in bidirectional charging in the long run, which is line with prior research [3]. Focusing on factors that drive the initial adoption of the technology, study 1 revealed that prior experience with electromobility goes along with an increased willingness to invest into bidirectional charging technologies. Further, our results highlight user preferences with regard to the design of specific business model aspects (i.e., preferred contractor, type of compensation, willingness to initially invest in the technology). Study 2 focused on designing app feedback using a gamification approach to motivate users in the long-rung which is especially important, as only if used continuously, the full advantaged of bidirectional charging technologies can be realized.

Based on our study results, we derive the following implications for practice: Concerning business model design, we find that respondents prefer to be able to choose between different providers for the certain components that are necessary for the system. This finding is especially true for participants who already own an electric car, which may be worthwhile to be considered in business models and tariff design for bidirectional charging. If potential customers already have experience with PEVs, they should be given the opportunity to choose from a variety of providers. For customers that show low experience with electromobility, it could be worthwhile to offer a combined bundle including all the necessary components to meet target group-specific preferences.

Further, our analyses on the respondents' preferences with regard to the type of compensation indicate, that revenues generated in the V2G use case should optimally be returned to the participants as savings in their electricity bill instead of a separate payout. Costumers thus seem to prefer reducing certain losses over receiving potential gains (i.e. they value savings in their electricity expenses higher than getting a separate payoff), a finding that is in line with classical literature on Prospect Theory[2]. Thus, emphasizing savings in electricity accounts should be given appropriate attention in the design of future business models.

Solely financial compensation will very likely not be sufficient to encourage user engagement on the long run. Especially for PEV-experienced customers increased integration of renewable energy as well as enhancing the overall grid stability (i.e., efficient energy use indicators) and social factors further represent relevant drivers for using bidirectional charging. In order to address these motivations, gamified feedback strategies can serve as a promising approach to create sustained user engagement.

Concerning gamified feedback design, study 2 showed that users' financial motivation is especially important to be addressed and should be considered first when implementing feedback: one way to do so is by allowing users to set themselves financial goals (i.e., earning points that can be exchanged for financial benefits). This implies that users can be motivated to constantly plug-in their cars as they strive for personal progress with regard to their behavior. Thus, they should be given detailed performance statistics in order to track and document their (financial) achievements. With regard to social feedback, our results indicate that building on (energy) communities and encouraging collaboration of users seems to be more important than competition among a group of users in bidirectional charging; individually contributing to a common goal represents a motivating factor. This could be further supported by additionally providing users with options to communicate, such as an in-app messenger. Finally, showing users their efficiency with regard to energy usage in a gamified manner might be a promising way to encourage regular bidirectional charging. Within study 2, we provided users with information concerning the share of energy and/or time exchanged charged in bidirectional vs. immediate charging. Those measures could be supported by further providing users with information on the effect of using bidirectional charging to achieve personal (e.g., increased autarchy level; environmental benefits through optimized private consumption) or grid-related (e.g., grid stability; efficient use of renewable energy) benefits.

In general, our studies focused on different user perspectives in the context of bidirectional charging. However, our findings are based on evaluations from people who have not yet used the technology themselves as only very few providers offer basic functionalities of bidirectional charging and field trials of the technology are scarce to date. Against this background, the BCM project offers a unique opportunity to assess our research with real users: within the field trial of the research project, 20 private users are currently testing the bidirectional charging technology since July 2021. This allows us to further investigate how real users perceive the business model and its components (e.g., in terms of satisfaction) as well as how gamified app feedback needs to be designed to create long-term engagement. Investigating real user behavior is especially worthwhile with regard to a successful implementation and market launch of the bidirectional charging technology. Focusing on not only

---

[2] Refers to the so-called *Endowment Effect* stating that consumers value objects that they already own higher, than objects that they are about to receive [14]. This effect appears due to consumers' *loss aversion*, which states, that potential losses are subjectively valued higher when being compared to potential gains of an objectively same amount [15].

technological developments but also user-related aspects becomes crucial, especially since the German government recently stated in their coalition agreement that *"Wir werden bidirektionales Laden ermöglichen"* (English: We will make bidirectional charging possible) [16, p. 52].

# References

1. BMW Group. https://www.press.bmwgroup.com/global/article/detail/T0338036EN/bidirectional-charging-management-bcm-pilot-project-enters-key-phase:-customer-test-vehicles-with-the-ability-to-give-back-green-energy?language=en. Accessed 4 Apr 2022
2. Sovacool, B.K., Axsen, J., Kempton, W.: The future promise of vehicle-to-grid (V2G) integration: a sociotechnical review and research agenda. Annu. Rev. Environ. Resour. **42**, 377–406 (2017)
3. Kämpfe, B., Zimmermann, J., Dreisbusch, M., Grimm, A.L., Schumann, J.H., Naujoks, F., Keinath, A., Krems, J.: Preferences and Perceptions of Bidirectional Charging from a Customer's Perspective–A Literature Review and Qualitative Approach. Electr. Mobility **2019**, 177–191 (2022)
4. Dreisbusch, M., Mang, S., Ried, S., Kellerer, F., Pfab, X.: Regulierung des netzdienlichen Ladens aus der Nutzerperspektive. ATZ-Automobil. Z. **122**(12), 68–73 (2020)
5. Noel, L., de Rubens, G.Z., Kester, J., Sovacool, B.K.: Navigating expert skepticism and consumer distrust: Rethinking the barriers to vehicle-to-grid (V2G) in the Nordic region. Transp. Policy **76**, 67–77 (2019)
6. Koivisto, J., Hamari, J.: The rise of motivational information systems: A review of gamification research. Int. J. Inf. Manage. **45**(4), 191–210 (2019)
7. Huotari, K., Hamari, J.: A definition for gamification: Anchoring gamification in the service marketing literature. Electron. Mark. **27**(1), 21–31 (2017)
8. Hamari, J., Koivisto, J., Sarsa, H.: Does gamification work? - a literature review of empirical studies on gamification. In: 2014 47th Hawaii International Conference on System Sciences, pp. 3025–3034. IEEE (2014)
9. Moreno-Munoz, A., Bellido-Outeirino, F.J., Siano, P., Gomez-Nieto, M.A.: Mobile social media for smart grids customer engagement: Emerging trends and challenges. Renew. Sustain. Energy Rev. **53**, 1611–1616 (2016)
10. Castri, R., Wemyss, D., Cellina, F., Luca, V. de, Frick, V., Lobsiger-Kaegi, E., Galbani Bianchi, P., Carabias, V.: Triggering electricity-saving through smart meters: Play, learn and interact using gamification and social comparison. In: Proceedings of the 1st ever Energy-Feedback Symposium – Teddinet. Edinburgh (UK) (2016)
11. ISO Homepage. https://www.iso.org/standard/25578.html. Accessed 5 Apr 2022
12. ISO Homepage. https://www.iso.org/obp/ui/#iso:std:iso:9241:-151:ed-1:v1:en. Accessed 5 Apr 2022
13. Hassenzahl, M., Burmester, M., Koller, F.: AttrakDiff: Ein Fragebogen zur Messung wahrgenommener hedonischer und pragmatischer Qualität. In: Mensch & Computer 2003, pp. 187–196. Vieweg + Teubner Verlag (2003)
14. Kahneman, D., Knetsch, J., Thaler, R.: Experimental tests of the endowment effect and the coase theorem. J. Polit. Econ. **98**(6), 1325–1348 (1990)
15. Kahneman, D., Tversky, A.: Prospect theory: An analysis of decision under risk. Econometrica **47**(2), 263–291 (1979)
16. SPD, Bündnis 90/Die Grünen, FDP: Mehr Fortschritt wagen. Bündnis für Freiheit, Gerechtigkeit und Nachhaltigkeit, Koalitionsvertrag 2021–2025 (2021)

# The Energy Transition in Germany

## Carbon Neutrality in the Balancing Act between Energy Demand and Energy Supply

Matthias Böger[✉] and Klaus Fuoss

Porsche Engineering Services GmbH, Bietigheim-Bissingen, Germany
{matthias.boeger,klaus.fuoss}@porsche-engineering.de

**Abstract.** Against the backdrop of climate change, Germany has committed to being carbon neutral by 2045. Even assuming a roughly 25% drop in energy consumption by then, it will still be necessary to close a gap of about 1300 TWh/year created by the move away from fossil energy sources. Although a full expansion of all renewable energy sources could theoretically lead to a completely self-sufficient and carbon-neutral supply in Germany, this is hardly conceivable in practice. This would require a massive transformation of the landscape as well as a corresponding acceptance among the population (for example, wind turbines in the vicinity of residential areas). The remaining alternatives are "green" energy imports in the form of hydrogen, eMethane, eFuels or the electricity directly.

Beyond merely obtaining the requisite amount of energy, availability in particular presents a second significant challenge. In general, renewable energy production is subject to volatility. The amount of energy that can be supplied depends on the respective environmental conditions (wind, sunlight) and is thus independent of the actual energy demand. In order to compensate for these fluctuations, energy storage systems of a corresponding size are indispensable. Currently, only synthetic, chemical energy carriers (hydrogen, eMethane, eFuels) can be used for large amounts of energy, such as those needed to store thermal energy for the winter.

**Keywords:** Energy storage · eFuel · Hydrogen · Renewable · Sustainable

## 1 Introduction

The Three Gorges Dam in China is the largest hydroelectric power plant in the world and is fed by the Yangtze River. It supplies the hydropower plant with around 30,000 $m^3$ of water per second and creates a reservoir some 660 km long and 1.5 km wide. Germany's biggest river, the Rhine, by contrast, carries only 3000 $m^3$ of water per second – a tenth of the volume carried by the Yangtze River. To meet Germany's current energy needs, it would take 24 power plants like the one operating at the Three

© Der/die Autor(en), exklusiv lizenziert an Springer Fachmedien Wiesbaden GmbH, ein Teil von Springer Nature 2023
A. Heintzel (Hrsg.): ATZLive 2022, Proceedings, S. 140–151, 2023.
https://doi.org/10.1007/978-3-658-41435-1_12

Gorges Dam in China. This figure alone shows that the shift to renewable energy cannot happen without a significant transformation of our landscape.

The projected global warming is a major challenge currently facing humanity. According to the view of the vast majority of climate researchers, temperatures are set to rise further. The expected consequences of global warming, some of which are already being observed today, include increasing weather extremes, growing drought zones, melting sea ice and glaciers, a rise in sea levels, and the thawing of permafrost soils with the release of methane hydrate.

Over the past 30 years, the amount of $CO_2$ emitted has changed significantly in regions around the world. While $CO_2$ emission-reduction efforts in Europe have been effective and emissions have been reduced by 28% [1], $CO_2$ emissions in other regions of the world have increased rapidly. Globally, $CO_2$ emissions have increased by 60% in the period since 1990; see Fig. 1.

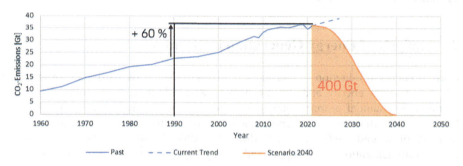

**Fig. 1.** $CO_2$ emissions per year in metric gigatons (Gt) per year (worldwide) and target profile [2]

The world is basically in a natural $CO_2$ cycle. The $CO_2$ emissions that occur in nature are broken down by equally natural processes. The annual global $CO_2$ emissions from the combustion of fossil energy carriers currently amount to around 36 metric gigatons (Gt) [3]. Although the environment partially absorbs this increase in the atmosphere through natural processes, an anthropogenic-based additional load of around 15 metric Gt $CO_2$/year remains in the atmosphere; see Fig. 2.

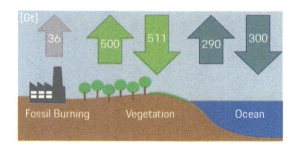

**Fig. 2.** Worldwide natural and anthropogenic $CO_2$ cycle per year [4]

142 M. Böger und K. Fuoss

To counter the effects of climate change, 197 countries around the world – including Germany – agreed as part of the Paris Agreement of 2015 to limit the rise in temperature due to the greenhouse effect to well below 2 °C above pre-industrial levels. To achieve this goal, no more than approximately 400 metric Gt of $CO_2$ should be emitted worldwide by the year 2040. With this amount, global warming can be limited to 1.5 °C with a probability of 67%, thus achieving the set target [5]. As concerns anthropogenic $CO_2$ emissions, only a few years remain to counteract climate change and to reduce $CO_2$ emissions accordingly. Against this backdrop, Germany has committed itself to being carbon-neutral by 2045.

In order to better assess the future boundary conditions and thus the market environment, Porsche Engineering examined the aspects of the energy transition for Germany in the context of this study and already defined the year 2040 as the year for achieving net climate neutrality; see Fig. 1. The aim was to assess the further development of the various possible drive technologies in conjunction with potential sustainable energy carriers.

## 2 Analysis of Energy Generation in Germany

Current energy production in Germany is mainly covered by hard coal, brown coal, oil, gas, nuclear energy and the renewable sources hydro, wind, solar energy and biomass. The annual primary energy demand amounts to more than 3200 TWh; see Fig. 3. This is offset by a final energy demand for electricity, heat, transportation and industry of around 2300 TWh. The difference (900 TWh) is attributable to energy losses from generation, local supply and energy exports. The use of fossil fuels causes $CO_2$ emissions of around 665 Mt/year in Germany [6].

A total of just over 500 TWh/year of the total final energy demand is currently covered by renewable sources. Assuming fully $CO_2$-neutral energy generation, this results in a gap of about 1800 TWh/year arising from the necessity of avoiding fossil fuels in the future. This gap can be closed from two directions: By saving energy on the one hand and expanding renewable energy sources or importing renewable energy carriers on the other. The following combination could be useful in this regard:

- Reduction of energy demand by 25%, which corresponds to the goal of the German Federal Environment Agency [7]
- Expansion of renewable energy generation or combination with energy imports

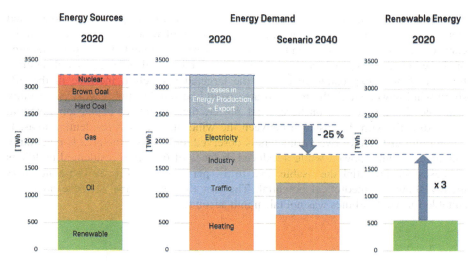

**Fig. 3.** Energy production and demand in Germany [8, 9]

## 3  Closing the Supply Gap

In general, the expansion possibilities of solar and wind power plants are limited by their space requirements, while biomass utilization is limited by the availability of basic materials. If studies with optimistic assumptions are used as a basis for the individual forms of generation, the following possible increase potentials result by 2040, calculated from today's level; see Fig. 4, [10–13]:

- Biomass: Factor of 1.5
- Solar energy: Factor of 5
- Wind offshore: Factor of 12
- Wind onshore: Factor of 6

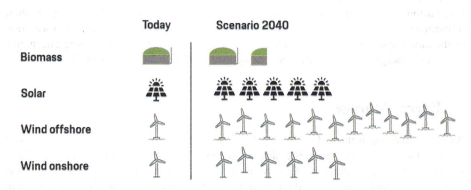

**Fig. 4.** Renewable energy expansion scenario for 2040

This forecast is based on the maximum use of all expansion options. The associated challenges are evident, for example, in the expansion of wind power capacities. In its current form, the North Sea wind farm [14] still has the potential to expand the area of the wind farm approximately fivefold. This includes all other approved, applied for and planned expansion areas. For the targeted twelvefold increase in the offshore wind power yield, the required area would be roughly twice what is currently available. In the case of onshore wind power, the theoretical expansion potential scales directly with the distance between the wind turbines and residential areas. If a minimum distance of two kilometers is observed, only very little area is potentially available for wind power, while a reduced distance of 600 m to residential buildings increases the potential the usable area to 14 per cent; see Fig. 5. These numbers merely show the theoretical potential. The fact that only some of these areas are suitable for wind turbines was not taken into account.

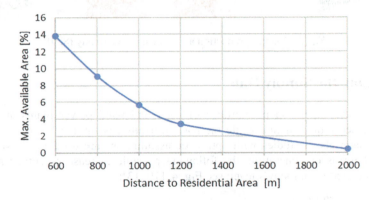

**Fig. 5.** Area potential for wind turbines in relation to the distance to residential areas [13]

If all renewable energy sources in Germany were fully developed, current knowledge suggests that Germany's energy needs could theoretically be met in a completely $CO_2$-neutral and self-sufficient manner. However, in view of widespread reservations among the population, for example relating to wind turbines in their immediate vicinity, the question arises as to whether the accompanying transformation of the landscape is socially desirable and politically feasible. The remaining alternative is "green" energy imports in the form of hydrogen, eMethane, eFuels or electricity.

## 4 Energy Storage

An additional requirement for closing the supply gap is the capacity to store generated energy. Generally speaking, the difficultly with renewable energies is their high volatility. The magnitude of this volatility was demonstrated in 2021, when the share of renewable energy production fell by 4.5% compared with the previous year, primarily because there was less wind on average over the course of the year. The maximum deliverable amount of energy is therefore not scalable in line with demand as is the case with conventional power plants, but rather dependent on the respective ambient conditions (wind, solar radiation). In the case of wind power, for example, the average annual capacity utilization in Germany is 22%; that is, five times as much power must be installed as is later generated on average. For solar power, the utilization rate over the year is 11 per cent. By comparison, the figure for brown coal is 46 per cent and for nuclear 86 per cent; these values are primarily limited not by their availability but by their demand [15, 16].

To align energy production and energy demand in time, intermediate storage of renewably produced energy is necessary. At times of low wind and little sunlight, that stored energy can then be tapped into. Fluctuations in energy production were divided into three categories: daily, weekly and seasonal.

From a daily perspective, volatilities in energy production result particularly from the absence of solar power during the hours of darkness. In times of high wind volume, on the other hand, the lower energy demand in the night hours can be used to store unused wind energy and make it available for use the next day; see Fig. 6. The storage facilities must be able to be replenished quickly and must also be able to make the energy available again quickly. The availability of the energy (fast conversion) is the priority here, with long-term storage capability more a secondary concern.

The availability of renewable energy also varies from week to week. Periods of a few days to weeks in which little or no wind blows or cloudy weather minimizes sunlight have to be bridged again and again. Especially in the summer months, wind conditions in Germany tend to be weaker, which has an accordingly unfavorable impact on energy production in this area. In terms of the storage facilities, the focus is on high storage capacity through high energy density and good storage capability of the energy carrier.

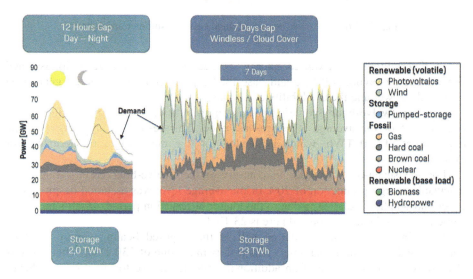

**Fig. 6.** Volatility of renewable energy sources on a daily and weekly basis using the example of electricity generation [17]

Further energy volatility arises from the heating of houses and apartments in winter. In this season, the energy demand is about twice as high as in the summer; see Fig. 7. Here again, large energy storage systems with high capacity and long-term storage capability are needed. The same applies to the strategic energy reserve, which aims to provide Germany with a self-sufficient supply of energy for at least three months. In addition to energy from renewable sources, fossil energy carriers remain suitable for this purpose. These then serve only as a fallback option and not as a source of energy for daily needs. It should be noted, however, that in this case power plant capacities with the corresponding infrastructure and personnel must continue to be kept available.

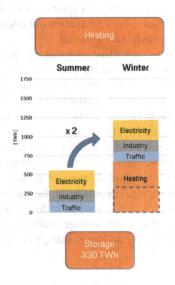

**Fig. 7.** Comparison of energy demand between summer and winter

In determining the required energy storage quantities, the scenario assumed general energy savings of 25% and base load-capable renewable energy generation from hydropower and the availability of biomass.

For the twelve-hour scenario, a remaining renewable energy production of five percent of the average level was assumed. This results in an energy storage requirement of 2.0 TWh, which is needed for a winter's night in Germany, for example.

For the seven-day scenario, in contrast to the twelve-hour case, a remaining renewable energy production of 20% of the average level was assumed despite cloud cover and windless conditions due to the longer observation period. On this basis, the amount of stored energy required here is 23 TWh.

For winter, it was assumed that half of the required heating energy must be obtained and stored in the summer. This results in a value of 330 TWh. This figure is also equivalent to the scale of an additional strategic reserve for electricity, industry and transport for three months.

# 5 Selection of the Energy Storage Technology

As the preceding consideration shows, the energy transition cannot be implemented in practice without appropriately dimensioned energy storage systems, at least in Germany, due to the high volatility. The importance of efficiency during feed-in and retrieval as well as storage is therefore secondary to the general storage capability in itself.

Selection of the correct storage type is based on the respective requirement in terms of response behavior and storage dimension: Short-term storage systems must be highly dynamic and ensure fast storage and availability. Long-term storage systems must combine a high energy density with good storage capabilities. Factors such as the required storage system size, the aging of the stored energy carrier and environmental protection (for example outgassing) are important factors here. Figure 8 shows a comparison of the currently technologically viable energy storage types for the three cases daily, weekly and seasonally from Figs. 6 and 7, assuming reconversion efficiency levels between 99% (battery) and 50% (synthetic chemical energy carriers).

| Storage Device | η* | Unit | Unit/Type of Measure | 12 Hours | 7 Days | Winter |
|---|---|---|---|---|---|---|
| Energy | | TWh | | 2.0 | 23 | 330 |
| Battery | 99 % | Number | 50 kWh** per Battery | 40 mil. | 460 mil. | 6.7 bil. |
| Pumped-storage | 95 % | Multiple | Current Capacity GER | 55 | 640 | 9,100 |
| Hydrogen | 50 % | Number | Cavern Storage | 19 | 220 | 3,100 |
| eMethane | 50 % | Number | Ø 60 m x 250 m, 100 bar | 6 | 71 | 1,000 |
| eFuel (Ethanol) | 50 % | Number | Cylindrical Tank Ø 40 m x 20 m | 27 | 310 | 4,500 |

\* Reconversion; \*\* 70 kWh, 70 % Residual capacity

**Fig. 8.** Comparison of energy storage types

The values shown indicate how many units of a certain storage medium are needed to have the required amount of energy, for example 23 TWh, available again as electricity after reconversion with the specified efficiency. Storage solutions that cannot be implemented to meet the requirements are highlighted in red. Solutions whose feasibility is questionable on this scale are colored yellow, while storage facilities marked green are suitable for the respective application and feasible in principle.

The viability of pumped-storage power plants is highly dependent on the geographical conditions. In Germany, it will therefore not be possible to multiply the currently available capacities to cover the energy demand even for twelve hours. Batteries and hydrogen, on the other hand, are suitable short-term storage systems for small amounts of energy. Batteries in particular are well suited to meet the required load or to store excess electricity. In addition to discarded batteries from old battery electric vehicles (BEVs), the BEV fleet itself could also serve as a buffer. The prerequisite in this case, however, is that the parked vehicles be connected to the grid as continuously as possible, which requires the corresponding infrastructure.

Due to its key position, hydrogen is becoming increasingly important both as a starting product for synthetic methane and synthetic fuels and as a substitute for fossil fuels in industrial applications. The storage capability poses a particular challenge here due to its volatility on the required scale. It remains to be seen which type of storage will prove most favorable here. Possible options include:

- Cavern storage, where the geological conditions are a decisive factor due to the requisite leaktightness [18]
- Large pressurized gas storage facilities
- The existing natural gas network [19]
- Novel storage approaches such as the thermal oil dibenzyltoluene [20].

Another possibility is to bind hydrogen to carbon and thus produce synthetic methane (eMethane) or a synthetic fuel (eFuel). These can be stored much more easily.

Due to its higher volumetric energy density and its lower volatility compared to hydrogen, eMethane can be stored particularly well in caverns. Caverns are natural cavities in the ground where natural gas, which consists of around 90% methane, has been deposited over the past millennia and was eventually extracted by humans. In Germany, around 300 such natural caverns already exist or are being developed and, together with pore storage facilities, currently have an energy storage capacity of 26.2 billion cubic meters, which corresponds to a heating energy of 240 TWh for e-methane and energy of approximately 120 TWh after reconversion into electricity [18, 21]. The calculation assumed an average cavern volume of 700,000 $m^3$, which corresponds to a cylinder with a diameter of 60 m and a height of 250 m. Furthermore, a storage pressure of 100 bar was assumed both for hydrogen and for eMethane.

A synthetic liquid fuel, on the other hand, can be stored in conventional fuel tanks such as those used at Frankfurt Airport.

Storing half of the thermal energy needed for winter (330 TWh after reconversion) would require either 4500 tanks for eFuel, 1000 caverns for eMethane, 3100 caverns for hydrogen, 9100 times the capacity of pumped-storage currently available in Germany, or 6.7 billion batteries, each with 50 kWh of usable storage capacity. As we can see, only eFuels and eMethane offer a feasible way to store energy for the winter. If the eMethane or eFuel is not converted back into electricity for heating, but the energy from combustion is used directly, the required storage quantities are also roughly halved. For the seven-day scenario, too, only hydrogen is added as a possible substitute for eFuel or eMethane.

In summary, only synthetic chemical energy carriers such as eMethane and eFuels are able to store the amounts of energy required to compensate for seasonal or weekly natural fluctuations in renewable energy production. Batteries, on the other hand, are only well suited to compensate for the daily fluctuations and to serve as a short-term energy buffer with high efficiency. Hydrogen plays an important role here, as it is needed as an intermediate medium. It can either be directly reconverted into electricity, stored for a short time, processed into eMethane or eFuel, or used as a carbon-neutral substitute product for industry. Hydrogen can be produced directly from electricity and water via electrolysis. Carbon is required to produce eMethane or eFuel. This is either taken directly from the air in the form of $CO_2$ or supplied via a closed $CO_2$ cycle. Since eFuels only release as much $CO_2$ during their combustion as was taken from the environment during their production and stored in them, they are virtually $CO_2$-neutral over the entire chain.

## 6 Future Energy Architecture

Figure 9 shows a possible architecture for energy generation, storage and use for the transport sector. The energy carriers already presented are used here again. For BEVs, the direct path from electricity production to the vehicle battery is the obvious choice for efficiency reasons. However, if the car cannot be charged directly, for example because there is neither enough wind nor sufficient sunshine, an intermediate storage system must be used. The previously stored energy can thus directly charge a BEV via a large-scale battery storage system. If a pumped-storage system is used, the energy must first be reconverted in order to charge the BEV. If the energy is in the form of hydrogen, it can be used directly in a fuel cell drive system. If eMethane or eFuels are available, they could also be used in internal combustion engines. In this case, it makes sense to use combustion engines that are designed for the use of these new synthetic energy carriers or hydrogen and have a correspondingly higher efficiency. It should be noted that these synthetic energy sources can, of course, also be converted back into electricity, so their energy is generally also available for BEVs. In this case, however, the efficiency after reconversion will differ only slightly from direct use in the internal combustion engine and will not generate any significant advantage.

In particular, it should be noted that although the energy carriers in Fig. 9 show a lower efficiency in production or reconversion in the progression to the right, they are more suitable in terms of their storage and thus also transport capability. Transportability is a relevant criterion, particularly in the case of energy imports.

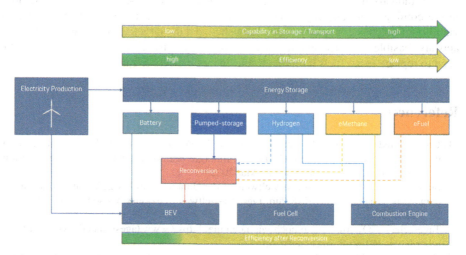

**Fig. 9.** Future architecture of energy generation, storage and use for the drive technology sector

# 7 Summary and Outlook

Against the backdrop of climate change, Germany has committed to being carbon neutral by 2045. Even assuming a roughly 25% drop in energy consumption by then, it will still be necessary to close a gap of about 1300 TWh/year created by the move away from fossil energy sources. Although a full expansion of all renewable energy sources could theoretically lead to a completely self-sufficient and carbon-neutral supply in Germany, this is hardly conceivable in practice. This would require a massive transformation of the landscape as well as a corresponding acceptance among the population (for example, wind turbines in the vicinity of residential areas). The remaining alternatives are "green" energy imports in the form of hydrogen, eMethane, eFuels or the electricity directly.

Beyond merely obtaining the requisite amount of energy, availability in particular presents a second significant challenge. In general, renewable energy production is subject to volatility. The amount of energy that can be supplied depends on the respective environmental conditions (wind, sunlight) and is thus independent of the actual energy demand. In order to compensate for these fluctuations, energy storage systems of a corresponding size are indispensable. Currently, only synthetic, chemical energy carriers (hydrogen, eMethane, e-fuels) can be used for large amounts of energy, such as those needed to store thermal energy for the winter.

The importance of efficiency in production will take a back seat to general storage capability.

Due to the range of different energy storage systems, $CO_2$-neutral energy sources will be available for BEVs, but also for vehicles with fuel cells or with combustion engines powered by hydrogen, eMethane or eFuels. Which drive technologies will ultimately be used will depend on the prevailing political and economic conditions. The global perspective will also play a decisive role here.

For industrialized countries, these developments will be the focus of attention in the coming years and represent a major challenge. On the basis of this study, the viability of this has essentially been established and the targeted implementation period appears feasible. The decisive factor here will be to raise awareness worldwide of the need to replace fossil fuels with renewable energy carriers as quickly as possible.

# References

1. Umweltbundesamt: Treibhausgas-Emissionen in der Europäischen Union. https://www.umweltbundesamt.de/daten/klima/treibhausgas-emissionen-in-der-europaeischen-union#hauptverursacher. Accessed 28 Mar 2022
2. Statista. (ed.).: $CO_2$-Emissionen weltweit in den Jahren 1960 bis 2020. https://de.statista.com/statistik/daten/studie/37187/umfrage/der-weltweite-co2-ausstoss-seit-1751/. Accessed 25 Mar 2022
3. Tagesschau: $CO_2$-Ausstoß so hoch wie noch nie. https://www.tagesschau.de/wirtschaft/co2-rekordhoch-101.html. Accessed 4 Apr 2022
4. Global Carbon Project: Global Carbon Budget 2021. https://www.globalcarbonproject.org/global/images/carbonbudget/Infographic_Emissions2021.pdf. Accessed 10 Mar 2022

The Energy Transition in Germany 151

5. IPCC, Climate Change 2021. https://www.ipcc.ch/report/ar6/wg1/downloads/report/IPCC_AR6_WGI_SPM_final.pdf#page=33. Accessed 4 Apr 2022
6. Statista (Hrsg.): Treibhausgasemissionen in Deutschland nach Sektoren des Klimaschutzgesetzes in den Jahren 1990 bis 2020 und Prognose für 2030. https://de.statista.com/statistik/daten/studie/1241046/umfrage/treibhausgasemissionen-in-deutschland-nach-sektor/. Accessed 4 Apr 2022
7. Umweltbundesamt, Energieeinsparpotenziale. https://www.umweltbundesamt.de/themen/klima-energie/energiesparen/energieeinsparpotenziale. Accessed 30 Mar 2022
8. Fachagentur Nachwachsende Rohstoffe e. V., Primärenergieverbrauch 2020. https://mediathek.fnr.de/grafiken/daten-und-fakten/bioenergie/primaerenergieverbrauch.html. Accessed 30 Mar 2022
9. Umweltbundesamt, Endenergieverbrauch 2020 nach Sektoren und Energieträgern. https://www.umweltbundesamt.de/bild/endenergieverbrauch-2020-nach-sektoren. Accessed 30 Mar 2022
10. Fachagentur Nachwachsende Rohstoffe: Biomasse-Potenziale. https://bioenergie.fnr.de/bioenergie/biomasse/biomasse-potenziale. Accessed 24 Nov 2020
11. WWF Deutschland: Zukunft Stromsystem II – Regionalisierung der erneuerbaren Stromerzeugung. https://www.oeko.de/fileadmin/oekodoc/Stromsystem-II-Regionalisierung-der-erneuerbaren-Stromerzeugung.pdf. Accessed 24 Nov 2020
12. Agentur für Erneuerbare Energien: Potenziale der Windenergie. https://www.unendlich-viel-energie.de/erneuerbare-energie/wind/onshore/potenziale-der-windenergie. Accessed 24 Nov 2020
13. Umweltbundesamt: Potential der Windenergie an Land. https://www.umweltbundesamt.de/sites/default/files/medien/378/publikationen/potenzial_der_windenergie.pdf. Accessed 28 Mar 2022
14. Wikimedia (Hrsg.): Offshore-Windkraftanlagen in der Deutschen Bucht. https://upload.wikimedia.org/wikipedia/commons/7/79/Karte_Offshore-Windkraftanlagen_in_der_Deutschen_Bucht.png. Accessed 25 Mar 2022
15. Fraunhofer-Institut für Solare Energiesysteme ISE: Installierte Netto-Leistung zur Stromerzeugung in Deutschland in 2020. https://energy-charts.info/charts/installed_power/chart.htm?l=de&c=DE&stacking=grouped&chartColumnSorting=default&year=2020, aufgerufen 2021/07/16
16. Fraunhofer-Institut für Solare Energiesysteme ISE: Jährliche Stromerzeugung in Deutschland 2020. https://energy-charts.info/charts/energy/chart.htm?l=de&c=DE&stacking=grouped&chartColumnSorting=default&interval=year&year=2020. Accessed 15 July 2021
17. Fraunhofer-Institut für Solare Energiesysteme ISE. https://energy-charts.info. Accessed 16 July 2021
18. Nationaler Wasserstoffrat der Bundesregierung, Die Rolle der Untergrund-Gasspeicher zur Entwicklung eines Wasserstoffmarktes in Deutschland. https://www.wasserstoffrat.de/fileadmin/wasserstoffrat/media/Dokumente/2022/2022-01-15_Positionspapier_H2-Speicher.pdf. Accessed 30 Mar 2022
19. Deutscher Verein des Gas- und Wasserfaches e. V.: Erstmals 20 Prozent Wasserstoff im deutschen Gasnetz. https://www.dvgw.de/der-dvgw/aktuelles/presse/presseinformationen/dvgw-presseinformation-vom-28102021-start-h2-beimischung-in-gasnetze. Accessed 4 Apr 2022
20. Stifterverband für die Deutsche Wissenschaft e. V., Geschäftsstelle Deutscher Zukunftspreis, Nominierung 2018. https://www.deutscher-zukunftspreis.de/de/team-3-2018. Accessed 30 Mar 2022
21. Erdöl Erdgas Kohle,137 Jg. 2021, Heft11, https://www.lbeg.niedersachsen.de/download/177852/Untertage-Gasspeicherung_in_Deutschland_Stand_1.1.2021_.pdf. Accessed 30 Mar 2022

# Autorenverzeichnis

**A**
Andresh, Manuel, 119

**B**
Benedikt, Florian, 86
Berg, Falko, 105
Bitsche, Otmar, 12
Böger, Matthias, 140
Bonifacio, Joao, 73

**D**
Danzer, Christoph, 33
Demmerer, Stephan, 21

**E**
Endres, Philip, 21

**F**
Fuoss, Klaus, 140

**G**
Geisler, Jürgen, 62
Gottorf, Simon, 46
Großgebauer, Uwe, 21
Gubin, Veronica, 86

**H**
Häge, Wolfgang, 73
Hara, Takafumi, 1
Hirsch, Sebastian, 130
Hofmann, Peter, 86
Hüther, Johannes, 62

**I**
Ito, Makoto, 1
Iwano, Ryuichiro, 1

**K**
Kellerer, Franziska, 130
Knappe, Pascal, 46
Konrad, Johannes, 86

**L**
Lensch-Franzen, Christian, 62

**M**
Mang, Stefan, 130
Müller, Stefan, 86

**P**
Paone, Alessio, 21
Passenberg, Benjamin, 12
Patyk, Andreas, 119
Pischinger, Stefan, 46
Poppitz, Alexander, 33
Prauße, Felix, 73
Prüger, Manfred, 33

**R**
Rempel, Thomas, 62
Ritzberger, Daniel, 105
Rosenfeld, Daniel Cenk, 86
Ruider, Martin, 21

**S**
Savelsberg, Rene, 46
Schäfers, Lukas, 46
Schenk, Alexander, 105
Schmalz, Mareike, 62
Schupp, Thomas, 73
Sens, Marc, 33
Späthe, Jana, 119
Sperber, Michael, 73
Suto, Tetsuya, 1

**T**
Takahashi, Akeshi, 1
Thewes, Matthias, 46

**V**
Varlese, Christian, 86
Voigt, Tobias, 33

© Der/die Herausgeber bzw. der/die Autor(en), exklusiv lizenziert an Springer
Fachmedien Wiesbaden GmbH, ein Teil von Springer Nature 2023
A. Heintzel (Hrsg.): ATZLive 2022, Proceedings, S. 153–154, 2023.
https://doi.org/10.1007/978-3-658-41435-1

## W

Wagner, Amalia, 62
Warth, Viktor, 73
Winkel, Matthias, 21

## Z

Zimmermann, Johanna, 130

9783658414344